calculus for business
and economics

calculus for business and economics

ROBERT L. CHILDRESS

school of business administration
university of southern california

Prentice-Hall, Inc., Englewood Cliffs, New Jersey

Printed in the United States of America

ISBN: 0-13-111492-1

Library of Congress Catalog Card Number: 78-164926

10 9 8 7 6 5 4 3 2 1

PRENTICE-HALL INTERNATIONAL, INC., *London*
PRENTICE-HALL OF AUSTRALIA, PTY. LTD., *Sydney*
PRENTICE-HALL OF CANADA, LTD., *Toronto*
PRENTICE-HALL OF INDIA PRIVATE LIMITED, *New Delhi*
PRENTICE-HALL OF JAPAN, INC., *Tokyo*

to carolyn

contents

appendix B
logarithms: laws of exponents **226**

tables

selected answers to odd-numbered questions **254**

preface

Calculus for Business and Economics is a first-level text in calculus and calculus-based business and economic models. The text is designed to satisfy the requirements of those schools that require an understanding of calculus and calculus-based models and those students who are interested in applying calculus to problems in economics, finance, production, marketing, and other business disciplines. The objective of the book is twofold. Calculus is introduced in a manner compatible with the background of most business and economics students. College algebra is the only prerequisite assumed for this text. Moreover, the theorem-proof approach to calculus is replaced by an explanation-example approach. Second, the text emphasizes business and economic application of calculus. Approximately 50 percent of the textual material is devoted to concepts and methodology with the remaining 50 percent concerned with applications and models.

The text consists of seven chapters. Five chapters cover differential calculus and applications, and two chapters cover integral calculus and applications. The book is designed for a one-semester course. Some teachers will prefer, however, to omit selected topics and condense the material in a manner such that linear programming, statistics, or other quantitative techniques can be taught during a portion of the semester. Since several excellent short texts on these subjects are currently available, topics other than calculus and calculus-based models are not discussed in this text.

Chapters 1 and 2 provide an introduction to the principles of differential calculus and examples of basic applications of differential calculus. Chapter 3 illustrates the use of calculus in revenue, profit, cost, and production models. Chapter 4 discusses multivariate and transcendental functions, including Lagrangian multipliers. Chapter 5 illustrates important multivariate and exponential business models. Chapter 6 introduces integral calculus and Chapter 7 applies integral calculus to business models.

Most individuals will wish to cover the concept in the order presented. Certain sections of the book may be omitted without loss of continuity. These sections are specified in the text.

Those individuals who have an understanding of calculus but have not been exposed to calculus-based business and economic models will find this text especially useful. An understanding of the principal applications of calculus to business and economics can be gained by studying the examples in Chapters 1, 2, 4, and 6 and studying Chapters 3, 5, and 7.

I am indebted to my associates at the University of Southern California, Professors Warren Erickson, Roger Glasser, Kirk Morrison, Ron Orr, Jack Pounders, Larry Press, Leonard Ross, Walter Ryder, and Michael White, for using the preliminary edition of the text and offering many helpful comments.

I would also like to thank Dr. John E. Freund, Arizona State University; Professor Richard M. Hesse, University of Southern California; Professor Fred E. Kindig, Ohio State University; and Professor Roman Weil, Jr., University of Chicago and Mr. George Kolodny for reviewing the text and for their suggestions.

calculus for business and economics

1

differential calculus

This text provides an introduction to calculus and some of the more important business models that are based upon calculus. Since an understanding of calculus requires an understanding of functional relationships between variables, our first task will be to introduce the concept of the function.

1.1 functions

A *function* is a mathematical relationship in which the values of a single dependent variable are determined from the values of one or more independent variables. To illustrate this definition, we will consider the *linear function*. The functional form of the linear function is

$$f(x) = a + bx. \tag{1.1}$$

$f(x)$ is the symbol, read f of x, that represents values of the dependent variable, and x is the symbol that represents values of the independent variable. The symbols a and b are termed the *parameters* of the function. a represents the value of the dependent variable when x is zero and b represents the slope or rate of change of the function. $f(x)$ varies according to the rule of the function as x varies. For the linear function, the rule of the function states that b is to be multiplied by x and this product added to a. This sum determines the value of the dependent variable $f(x)$. Since the value of $f(x)$ depends upon the value of x, $f(x)$ is called the dependent variable and x is termed the independent variable. The term *variable* refers to a quantity that is allowed to assume different numerical values.

An example of a linear function is

$$f(x) = 2 + 1x.$$

This function is graphed in Fig. 1.1. $f(x)$ is the dependent variable in Fig. 1.1 and x is the independent variable. The parameters are $a = 2$ and $b = 1$. The rule of the function states that $f(x)$ is equal to 2 plus $1x$. The table in Fig. 1.1 gives six possible values of x and $f(x)$.

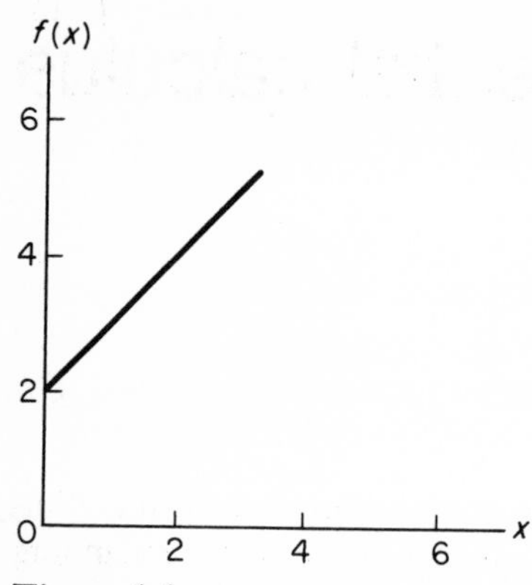

x	0	1	2	3	4	5
$f(x)$	2	3	4	5	6	7

Figure 1.1

The mathematical term for the permissible values of the independent variable is the *domain* of the function. If, for example, the analyst is interested only in positive values of the independent variable, he would state that the domain of x consists of all positive numbers. If he were concerned with only positive integer values of x, then the domain of x would consist of only positive integers.

The permissible values of the dependent variable are termed the *range of the function*. The range of the function is those values that the dependent variable assumes as the independent variable takes on all values in the domain. In the example illustrated in Fig. 1.1, if the domain of the function is all values of x between $x = 0$ and $x = 5$, then the range contains all the numbers between $f(x) = 2$ and $f(x) = 7$. Similarly, if the domain of the function is defined as all positive values of x, then the range of the function is all positive values of $f(x)$ between $f(x) = 2$ and $f(x) = \infty$. The concepts of functions and the domain and range of the function are illustrated in the following examples.

Example. Harvey West, an analyst for Pacific Soft-Drink Company, is attempting to develop a cost function for a diet cola. From the Accounting Department he has learned that fixed costs are \$10,000 and variable costs are \$0.03 per bottle. Present capacity limitations are 50,000 bottles per month, or 600,000 bottles per year. Mr. West wishes to develop the cost function and to specify the domain and range of the function.

We shall represent the number of bottles of the diet cola by x and the total cost by $f(x)$. Since total cost is composed of fixed and variable costs, the cost function on a yearly basis is

$$f(x) = 10{,}000 + 0.03x.$$

The domain of the function is $x = 0$ to $x = 600{,}000$. The corresponding range is $f(x) = \$10{,}000$ to $f(x) = \$28{,}000$.

Example. Bill Short owns a small auto repair shop. Mr. Short acts as supervisor and employs three auto repairmen. His shop is operated on a 40-hour-per-week basis, and his employees are guaranteed 40 hours of pay each week. Weekly labor costs consist of a fixed component and a sum that varies with the amount of overtime during the week. The weekly salaries of Mr. Short and the three repairmen total \$750. The overtime rate is \$5.00 per hour for the employees. Mr. Short does not draw overtime pay. The maximum amount of overtime per employee is 20 hours per week. Determine the cost function and the domain and range of the function.

The cost function consists of the sum of the weekly salaries and the overtime pay. If we represent the overtime hours per week by x and the total weekly salary by $f(x)$, then the cost function is

$$f(x) = 750 + 5x.$$

The domain of the function is $x = 0$ to $x = 60$, and the corresponding range of the function is $f(x) = \$750$ to $f(x) = \$1050$.

Example. Bud Brewer is attempting to predict sales for his company for the coming year. Sales for the past five years are plotted on the graph shown below. On the basis of his knowledge of the market and the historical data, Bud believes that a linear function which passes through the data points for year 1 and year 5 will provide a reasonable forecast for sales for the coming year. Determine the function along with the range and domain of the function.

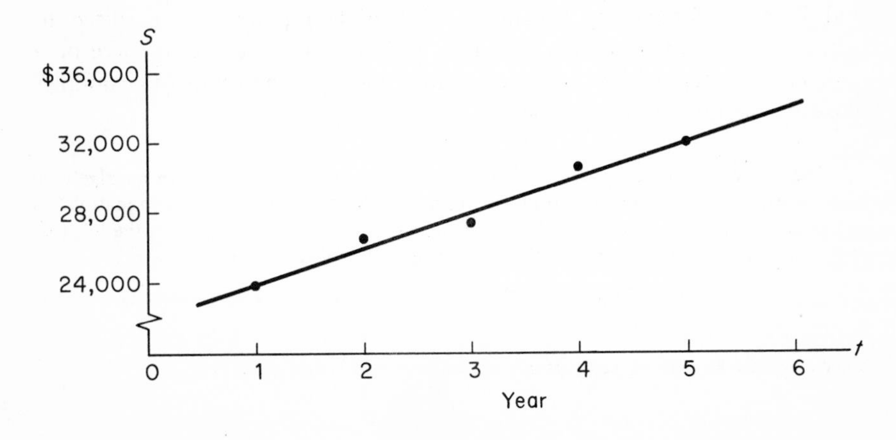

The method of establishing functions through data points is described in Appendix A. Two data points are required to establish a linear function. From Bud's statements, these data points are ($t = 1$, $S = \$24{,}000$) and ($t = 5$, $S = \$32{,}000$). Substituting these values in the general form of the linear function, $S(t) = a + bt$, gives the two equations

$$24{,}000 = a + b(1)$$
$$32{,}000 = a + b(5),$$

which when solved simultaneously for a and b give $a = 22{,}000$, $b = 2000$, and

$$S(t) = \$22{,}000 + \$2000t.\dagger$$

Since Bud Brewer is interested in predicting sales based upon this function for the coming year, year 6, he has implied that the domain of the function is $x = 1$ to $x = 6$. The corresponding range of the function is $f(1) = \$24{,}000$ to $f(6) = \$34{,}000$. The forecast for year 6 is $f(6) = \$34{,}000$. In the examples, the values of x consisted of a finite number of points. For example, the domain of the cost function for Pacific Soft-Drink Company was the 600,001 integer numbers between $x = 0$ and $x = 600{,}000$. In cases such as this, it is acceptable for purposes of analysis to assume that the domain and the range are continuous rather than discrete. We will make this assumption in this text.

The simple linear function is one of the more useful functions in business and economics. Another function that has important applications is the *quadratic function*. The quadratic function has the general form

$$f(x) = a + bx + cx^2. \tag{1.2}$$

$f(x)$ is again the dependent variable, x the independent variable, and a, b, and c are the parameters of the function. As an example of a quadratic function, consider the function

$$f(x) = 7.684 - 1.912x + 0.228x^2.$$

This function is plotted in Fig. A.2 of Appendix A. We remember from Bud Brewer's forecasting function that two data points are required to define a linear function. It is shown in A.2 of Appendix A that three data points are necessary to define a quadratic function. The following examples illustrate the quadratic function.

---

†The two equations can be solved for a and b by remembering that in an algebraic equation equal values can be subtracted from both sides of the algebraic equation. In this example 24,000 is equal to $a + b(1)$. Thus we can subtract 24,000 from the left side of the second equation and $a + b(1)$ from the right side of the second equation.

$$32{,}000 - 24{,}000 = a + b(5) - a - b(1)$$
$$8000 = 4b, \qquad b = 2000.$$

Since $a = 24{,}000 - b$, we see that $a = 22{,}000$.

Example. Bill Cook owns a small machine shop. Because of the limited capacity of the shop, average costs are related to the volume of output. Cost studies undertaken by Bill show that the average cost of producing 90 units of output is $850.00 per unit, the average cost of producing 100 units is $800.00 per unit, and the average cost of producing 120 units is $825.00 per unit. Determine the functional relationship between average cost and output.

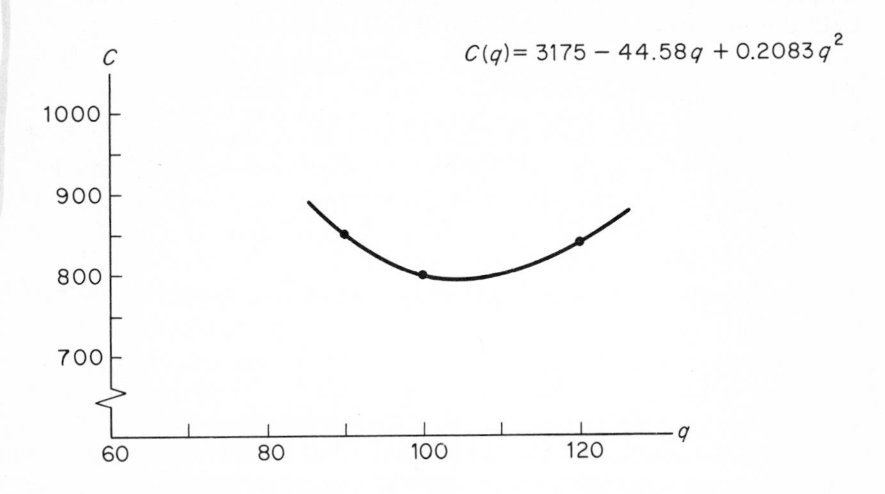

A plot of the three data points shows that a function which decreases and then increases describes the cost-output data. This type of functional relationship is described by the quadratic function.† The procedure for determining the function is described in Sec. A.2 of Appendix A. The data points are substituted into the general form of the quadratic function, $f(q) = a + bq + cq^2$, and the resulting three equations are solved simultaneously for the parameters a, b, and c. The three equations are

$$\begin{aligned} 850 &= a + b(90) + c(90)^2 \\ 800 &= a + b(100) + c(100)^2 \\ 825 &= a + b(120) + c(120)^2. \end{aligned}$$

The solution of the three simultaneous equations is $a = 3175$, $b = -44.58$, and $c = 0.2083$. The average cost function is

$$C(q) = 3175 - 44.58(q) + 0.2083(q)^2.$$

---

†The sign and magnitude of the parameters b and c determines the general shape of the quadratic function. If c is positive and b is negative, the function has the general shape shown in the Bill Cook example. If c is negative and b is positive, the function has the general shape shown in the Fred Mason example. The magnitude of the parameter a determines the vertical position of the function.

Example. Fred Mason is concerned with the effect of fatigue on the productivity of skilled machinists. He has observed that productivity, which he defines as standard units of output per man-hour, increases as the number of hours that an individual works increases up to approximately 40 hours per week and then begins to decline as fatigue becomes a factor. Mason believes that three representative observations of the effect of man-hours on productivity are 9.3 units of output during the 35th hour of work, 9.5 units during the 40th hour, and 8.5 units during the 45th hour.

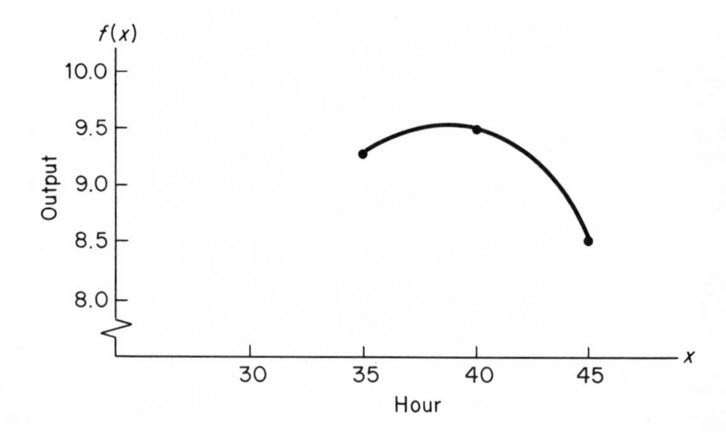

The observations are plotted below. Productivity as a function of the hour of work increases and then decreases. This type of a relationship illustrates the second general shape of the quadratic function. Again referring to A.2 of Appendix A, the data points are substituted in the general form of the quadratic function.

$$\begin{aligned} 9.3 &= a + b(35) + c(35)^2 \\ 9.5 &= a + b(40) + c(40)^2 \\ 8.5 &= a + b(45) + c(45)^2. \end{aligned}$$

Solving the three equations simultaneously for a, b, and c gives $a = -25.7$, $b = 1.84$, and $c = -0.024$. The productivity function is

$$f(x) = -25.7 + 1.84x - 0.024x^2 \qquad \text{for } 35 \leq x \leq 45.$$

Linear and quadratic functions belong to a class of functions termed *polynomial functions*. Polynomial functions are defined in Sec. 1.2.3. Another example of a polynomial function is the *cubic function*. The cubic function has the general form

$$f(x) = a + bx + cx^2 + dx^3. \tag{1.3}$$

As an illustration of the cubic function, the function $f(x) = 5 + 2.5x - x^2 + 0.1x^3$ is plotted in Fig. 1.2. $f(x)$ is again the dependent variable and x the independent variable.

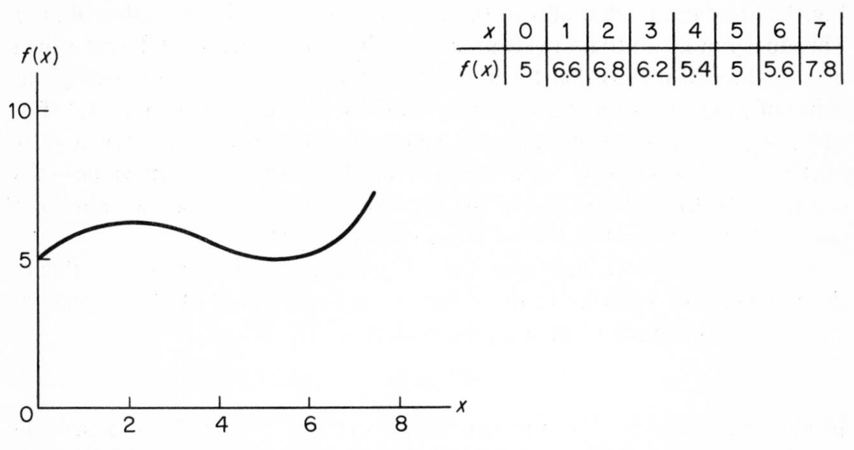

x	0	1	2	3	4	5	6	7
$f(x)$	5	6.6	6.8	6.2	5.4	5	5.6	7.8

Figure 1.2

We have stated that a function has only one dependent variable but may have one or more independent variables. Functions in which the single dependent variable is related to more than one independent variable are termed multivariate functions. An example of a multivariate function is

$$f(x_1, x_2) = 2x_1 + 3x_1x_2 + 6x_2,$$

where $f(x_1, x_2)$ is the dependent variable, x_1 is an independent variable, and x_2 is a second independent variable. These functions are difficult to plot on two-dimensional paper, since they require one axis (or dimension) for each variable. A function with two independent variables and one dependent variable requires three dimensions for plotting. A multivariate function with three independent variables cannot be plotted, since four dimensions are required and only three dimensions are available. In spite of the fact that multivariate functions with more than two independent variables cannot be plotted, they are quite necessary to describe many business and economic models. An example of a multivariate function with two independent variables follows. Multivariate functions are considered in detail in Chapter 4.

Example. Sales of the Southern Distributing Company consist of sales made through retail and wholesale outlets. Profit per dollar of retail sales is \$0.15 and profit per dollar of wholesale sales is \$0.05. If we represent retail sales by x_1 and wholesale sales by x_2, the profit from sales is

$$P(x_1, x_2) = 0.15x_1 + 0.05x_2.$$

A distinguishing characteristic of a function (of either one or more than one independent variable) is that for each value of the independent variable or combination of values of the independent variables, there exists only one value of the dependent variable. As an example, consider the function in Fig. 1.1. For each value of x in the domain, there is only one value of $f(x)$. This does not mean that $f(x)$ cannot have the same value for different values of x. Rather, it means that for a specific value of x there can exist only one value of $f(x)$. As a second example, consider the function in Fig. 1.2. For each value of x there is only one value of $f(x)$. However, when $x = 0$, $f(x) = 5$ and when $x = 5$, $f(x) = 5$. This shows that different values of x can result in the same values of the dependent variable. A single value of x cannot, however, lead to two or more values of $f(x)$.

There are, however, mathematical relationships in which a single value of the independent variable leads to more than one value of the dependent variable. An example of such a relationship is

$$y^2 = x.$$

In this relationship, y is the dependent variable and x the independent variable. A graph of $y^2 = x$ is given in Fig. 1.3. The domain of the relationship in this graph is $x = 0$ to $x = 9$.

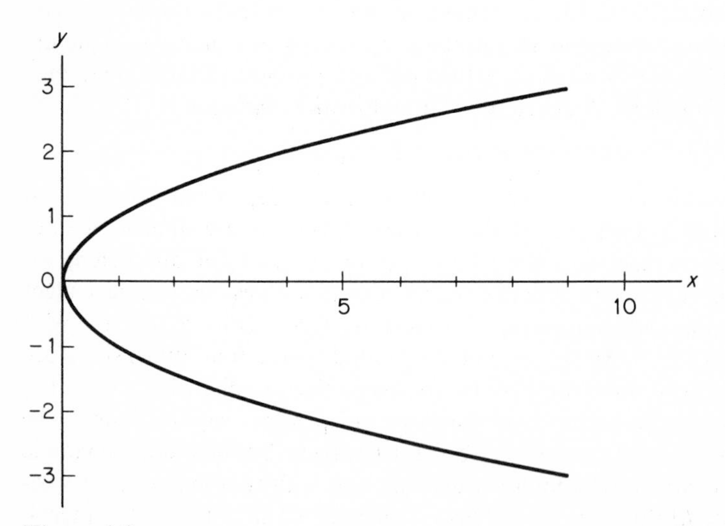

Figure 1.3

If $y^2 = x$, then $y = \pm\sqrt{x}$ and for each value of x, there are two values of y. This violates the requirement that there exists only one value of the dependent variable for each value or combination of values of the independent variables. Consequently, $y = \pm\sqrt{x}$ is not a function. Relationships between variables in which two or more values of the dependent variable exist for each value or combination of values of the independent variables are termed *relations*. The mathematics of relations is not considered in this text.

Before we can define the meaning of differential calculus, we must define the concepts of rate of change, limit, and continuity. These concepts, along with the concept of a function, are fundamental to an understanding of differential calculus.

1.1.1 RATE OF CHANGE

The rate of change of a function is defined as the change in the value of the dependent variable divided by the change in the value of the independent variable. The formula for the rate of change of a linear function can be developed from the definition of rate of change. The functional form of the linear function was given by Formula (1.1) as

$$f(x) = a + bx. \tag{1.1}$$

If the independent variable x is increased by the arbitrary amount Δx, the value of the independent variable becomes $(x + \Delta x)$. Replacing x by $(x + \Delta x)$, we obtain

$$f(x + \Delta x) = a + b(x + \Delta x). \tag{1.1'}$$

To determine the increase in the dependent variable due to the increase Δx in the independent variable, we subtract (1.1) from (1.1′).

$$\begin{aligned} f(x + \Delta x) - f(x) &= (a + b(x + \Delta x)) - (a + bx) \\ f(x + \Delta x) - f(x) &= a + bx + b\,\Delta x - a - bx \\ f(x + \Delta x) - f(x) &= b\,\Delta x. \end{aligned}$$

Remembering that the rate of change of a linear function is defined as the change in the value of the function divided by the change in the value of the independent variable, we see that

$$\frac{f(x + \Delta x) - f(x)}{\Delta x} = \frac{b\,\Delta x}{\Delta x} = b,$$

which is the slope of the linear function. Using the symbol Δ to represent change, we can rewrite the formula for the rate of change of a linear function as

$$\text{Rate of change} = \frac{\Delta f(x)}{\Delta x} = \frac{f(x + \Delta x) - f(x)}{\Delta x}. \tag{1.4}$$

In deriving the formula for the slope or rate of change of the linear function, we introduced the symbol Δ to represent change. The symbol Δx refers to an incremental quantity of the analyst's choosing. For example, if $x = 2.0$

and $\Delta x = 0.1$, then $x + \Delta x = 2.1$. The symbol $\Delta f(x)$ represents the change in $f(x)$ which is associated with the change in x, that is,

$$\Delta f(x) = f(x + \Delta x) - f(x).$$

The formula for the slope or rate of change of a linear function is illustrated by the following examples.

Example. $f(x) = -2x + 1$.

$$\begin{aligned}
\text{Rate of change} &= \frac{\Delta f(x)}{\Delta x} = \frac{f(x + \Delta x) - f(x)}{\Delta x} \\
&= \frac{-2(x + \Delta x) + 1 - (-2x + 1)}{\Delta x} \\
&= \frac{-2x - 2\,\Delta x + 1 + 2x - 1}{\Delta x} = \frac{-2\,\Delta x}{\Delta x} = -2.
\end{aligned}$$

Example. $f(x) = \dfrac{x - 2}{3}$.

$$\begin{aligned}
\text{Rate of change} &= \frac{\Delta f(x)}{\Delta x} = \frac{f(x + \Delta x) - f(x)}{\Delta x} \\
&= \frac{\dfrac{(x + \Delta x) - 2}{3} - \dfrac{(x - 2)}{3}}{\Delta x} = \frac{(x + \Delta x) - 2 - (x - 2)}{3\,\Delta x} \\
&= \frac{x + \Delta x - 2 - x + 2}{3\,\Delta x} = \frac{\Delta x}{3\,\Delta x} = \frac{1}{3}.
\end{aligned}$$

1.1.2 LIMIT OF A FUNCTION

The existence of a limiting value of a function for certain types of functions can easily be demonstrated. As an example, consider the function

$$y = 1 + (\tfrac{1}{2})^x \qquad \text{for } x = 0, 1, 2, 3, 4, \ldots, \infty.$$

As x approaches infinity, written $x \to \infty$, y assumes the values

$$2,\ 1\tfrac{1}{2},\ 1\tfrac{1}{4},\ 1\tfrac{1}{8},\ 1\tfrac{1}{16},\ 1\tfrac{1}{36},\ 1\tfrac{1}{64},\ \ldots.$$

It is seen that as $x \to \infty$, $y \to 1$; i.e., the limiting value of y as x approaches infinity is 1. Mathematically, this is expressed as

$$\operatorname*{limit}_{x \to \infty} y = \operatorname*{limit}_{x \to \infty} [1 + (\tfrac{1}{2})^x] = 1.$$

A limit can be defined as follows. Let $f(x)$ be a function and a some fixed number. If $f(x)$ approaches some fixed number L as x approaches a, then L is termed the limit of the function $f(x)$ as x approaches a. This is expressed as

$$\operatorname*{limit}_{x \to a} f(x) = L. \tag{1.5}$$

The selection of a determines the value of the limit. Given the simple linear function,

$$y = x + 2,$$

we see that the limit of y as x approaches 3 is 5.

$$\lim_{x \to 3} y = \lim_{x \to 3} (x + 2) = 5.$$

The limit of 5 as x approaches 3 means that as x comes closer and closer to 3, y comes closer and closer to 5. However, for the same function, the limit as x approaches infinity does not exist. That is, there is no fixed number L that y approaches as x approaches infinity. This function, therefore, has no limiting value as x approaches infinity. This illustrates the fact that a function which has a limiting value for certain values of the independent variable need not have a limiting value for all possible values of the independent variable.

The following examples illustrate limiting values of selected functions.

Example. $\lim_{x \to \infty} \left(\frac{1}{x + 1}\right) = 0.$

Example. $\lim_{x \to 6} (x^2) = 36.$

Example. $\lim_{x \to 2} \left(\frac{x^2 - 4}{x - 2}\right) = \lim_{x \to 2} \left(\frac{(x + 2)(x - 2)}{(x - 2)}\right) = \lim_{x \to 2} (x + 2) = 4.$

It is tempting in determining limits simply to set $x = a$ and determine $f(a)$. The preceding example illustrates, however, that this is not the proper procedure. If we set $x = 2$, we are attempting to divide $(x^2 - 4)$ by 0. *Division by 0 is not permitted.* The correct procedure is to eliminate the term $(x - 2)$ from the denominator by division of the denominator into the numerator. This results in $L = 4$ rather than the mathematically undefined term $(\frac{0}{0})$.

Figure 1.4 illustrates several possible situations regarding the limit of $f(x)$ as x approaches a.

Diagrams (a) and (b) both show functions that have the limit L as x approaches a. Diagrams (c) and (d) illustrate functions for which limits do not exist as x approaches a. In diagram (c), the limit as x approaches a from the left is L_1, whereas the limit as x approaches a from the right is L_2. An additional requirement for a limit is that the left side and right side limits must be the same. This is not the case in diagram (c). Consequently, the function does not have a limit as x approaches a.

Diagram (d) shows a function that has a limit for all values of x except as x approaches 0. As x approaches 0, the function $f(x)$ approaches $+\infty$ or

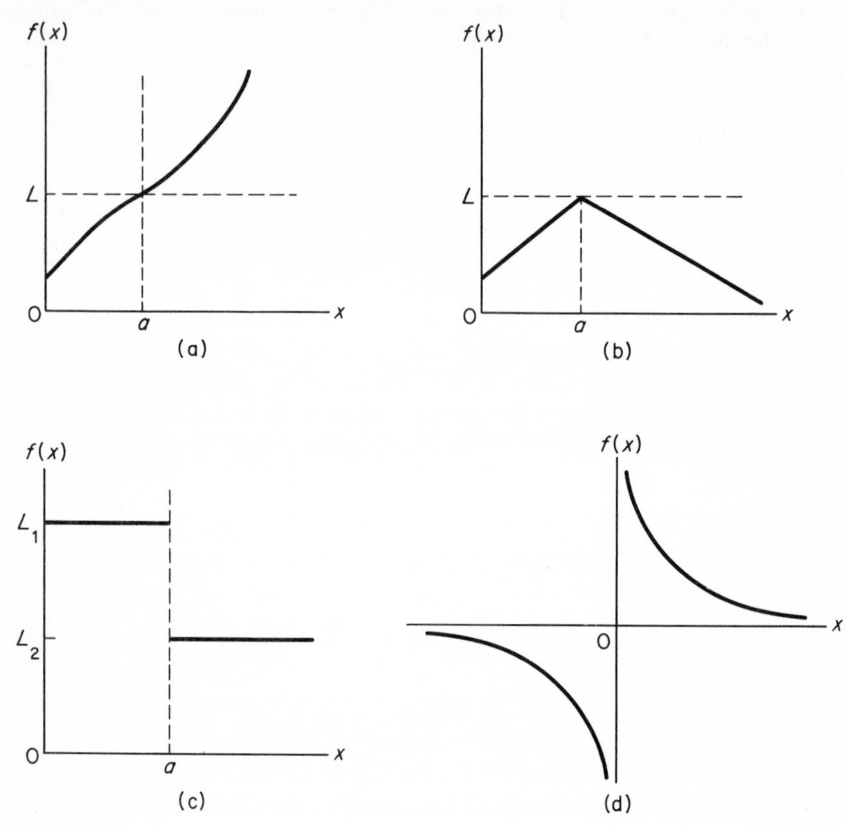

Figure 1.4

$-\infty$. Infinity is not a limiting value of the function; i.e., ∞ is not considered as a limit. Consequently, the function has no limit as x approaches 0. The function in diagram (d) does, however, have a limit as x approaches any value other than 0.

1.1.3 FORMAL DEFINITION OF LIMIT

A function $f(x)$ has a limit L as x approaches a when, as x is given in a sequence of values approaching a from either the left or the right, the corresponding value of $f(x)$ can be made to approach the constant L as closely as desired. This definition is illustrated in Fig. 1.5. Let $[L - f(x - \Delta x)] = f_1$ and $[f(x + \Delta x) - L] = f_2$. Note that f_1 or f_2 can be made as close to 0 as desired by permitting Δx to approach 0. As f_1 and f_2 approach 0, $f(x)$ approaches the constant L. This illustrates the requirement in the definition that $f(x)$ can be made to approach the constant L as closely as desired.

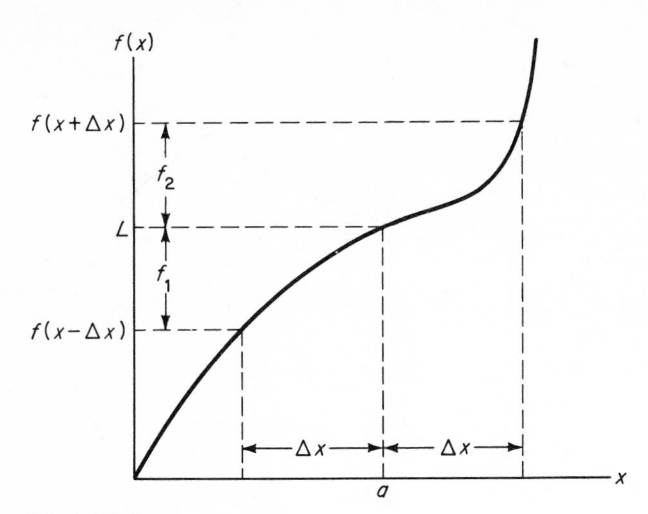

Figure 1.5.

1.1.4 CONTINUITY

An understanding of the concept of continuity is necessary to understand differential calculus. Simply stated, a function is continuous in an interval from $x = b$ to $x = c$ if the function has no breaks or jumps in the interval. Fig. 1.6 shows examples of both continuous and discontinuous functions.

The function in diagram (a) is continuous. The function in diagram (b) is also continuous even though it reaches a point at $x = a$. There is a break or step in the function depicted in diagram (c), which causes a discontinuity at $x = a$. Similarly, the function in diagram (d) has a break at $x = a$ and is therefore discontinuous at $x = a$.

The requirements for continuity of a function at a point $x = a$ are

1. $f(a)$ is defined.
2. $\lim_{x \to a} f(x)$ exists.
3. $\lim_{x \to a} f(x) = f(a)$, whether x approaches a from the left or the right.

The following examples illustrate the requirements for continuity.

Example. $f(x) = x + 4$, for $0 \leq x \leq 4$. Specify the continuity at $x = 6$. The function is not defined at $x = 6$. Consequently, the function is discontinuous at that point.

Example. $f(x) = \dfrac{10}{x - 2}$ for $0 \leq x \leq \infty$. Specify the continuity at $x = 2$. The limit of $f(x)$ as x approaches 2 does not exist. Therefore, the function is discontinuous at $x = 2$.

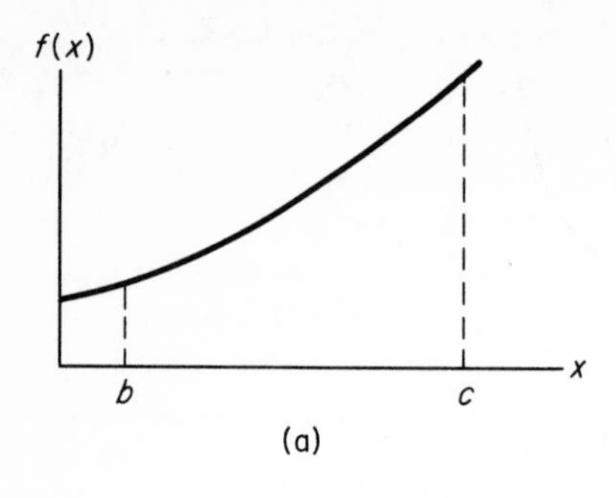

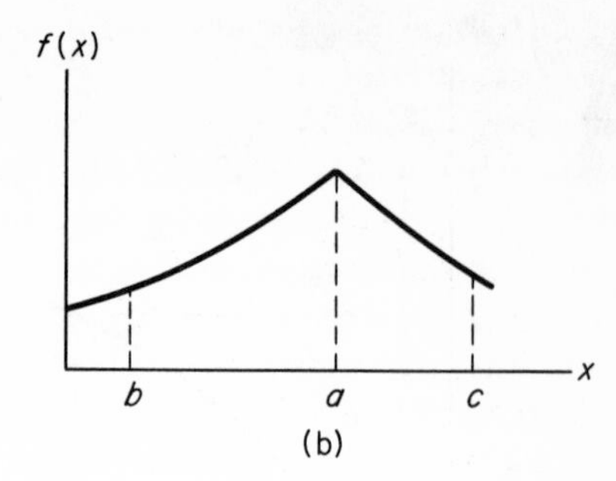

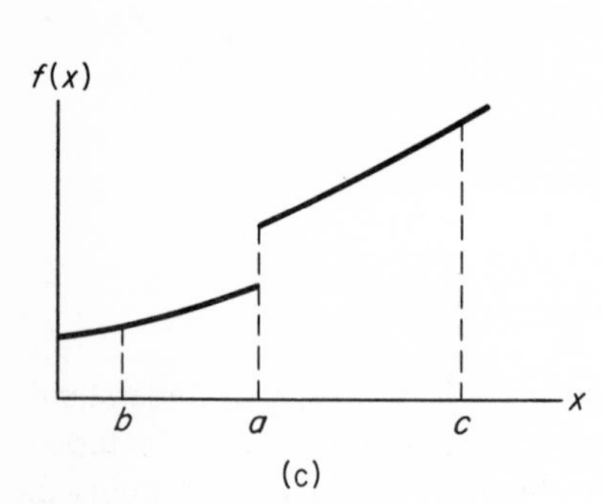

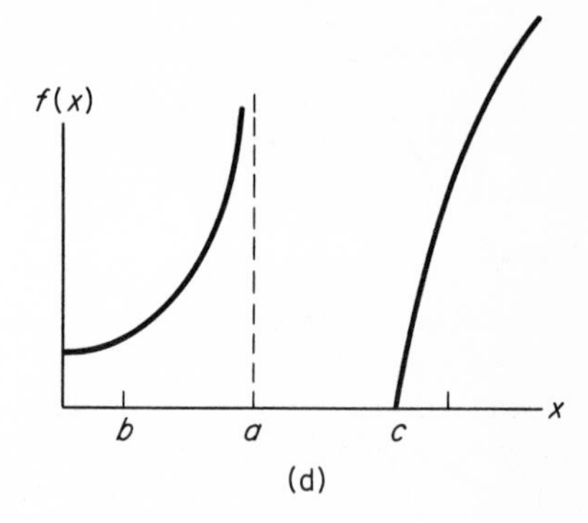

Figure 1.6

Example. $f(x) = x^2$, for $0 \leq x \leq \infty$. Specify the continuity at $x = 6$. $f(6)$ is defined and the limit as x approaches 6 exists. Furthermore, the limit as x approaches 6 from the left equals the limit as x approaches 6 from the right. Therefore, the function is continuous at $x = 6$.

1.1.5 INSTANTANEOUS RATE OF CHANGE

The instantaneous rate of change of the function $y = f(x)$ is defined as the limit of the change in the dependent variable divided by the change in the independent variable. The instantaneous rate of change differs from the rate of change as previously defined by Formula (1.4) in that the magnitude of Δx is not specified in Formula (1.4). The magnitude of Δx approaches 0 for the instantaneous rate of change. The formula for the instantaneous rate of change is

$$f'(x) = \operatorname*{limit}_{\Delta x \to 0} \frac{\Delta f(x)}{\Delta x};$$

or alternatively,

$$f'(x) = \operatorname*{limit}_{\Delta x \to 0} \frac{f(x + \Delta x) - f(x)}{\Delta x}, \tag{1.6}$$

where $f'(x)$ is a symbol representing the instantaneous rate of change.

The purpose of the instantaneous rate of change is illustrated by Fig. 1.7. The instantaneous rate of change is equivalent to the slope of a function for an arbitrarily chosen value of the independent variable. Both slope and instantaneous rate of change are defined as the change in the value of the dependent variable divided by the magnitude of the change in the independent variable. The slope of the linear function in Fig. 1.7(a) is constant for any arbitrary Δx. The slope of the curvilinear function in Fig. 1.7(b) depends, however, upon the selection of Δx. If the objective is to determine the slope of the function, or, alternatively, the instantaneous rate of change of the function at $x = a$, selection of a large interval for Δx provides a poor approximation of the true slope for the curvilinear function. Consequently, Δx must be allowed to approach 0 in order to determine the slope of the function at $x = a$.

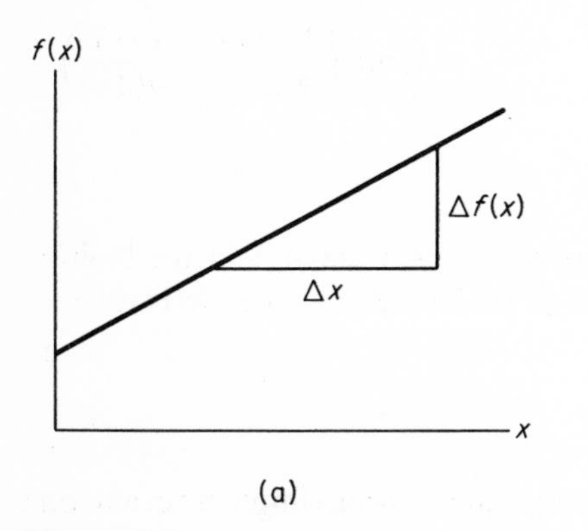

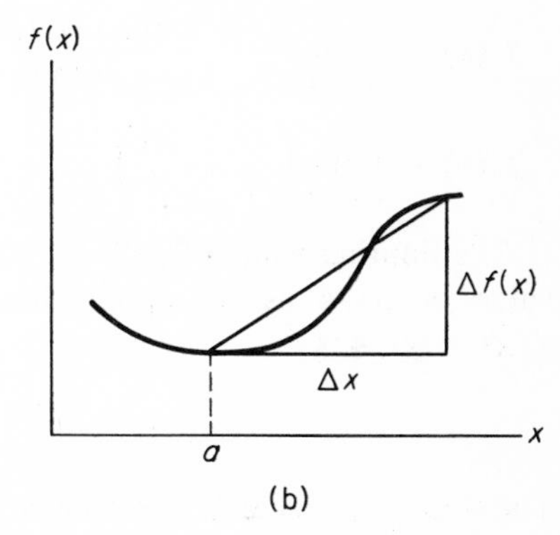

Figure 1.7

1.2 the derivative

We can now introduce the concept of the derivative. The derivative of a function is defined as the *instantaneous* rate of change of the function. The formula for the derivative is

$$f'(x) = \underset{\Delta x \to 0}{\text{limit}} \frac{\Delta f(x)}{\Delta x};$$

or alternatively,

$$f'(x) = \underset{\Delta x \to 0}{\text{limit}} \frac{f(x + \Delta x) - f(x)}{\Delta x}, \tag{1.6}$$

where $f'(x)$ is a symbol representing the derivative of the function $f(x)$.

The derivative provides a method for determining the slope or rate of change of a curvilinear function at a value of the independent variable, $x = a$. As an example, the function $f(x) = x^2/4$ is plotted in Fig. 1.8. Assume that we wish to determine the slope of this function for different values of the independent variable. Applying Formula (1.4) for the slope of a linear function, we can determine the average slope of the function between two values of the independent variable. Thus, between $x = 2$ and $x = 4$, the average slope is $\Delta f(x)/\Delta x = 3/2 = 1.5$ (see Fig. 1.8). Note that there is no single slope in this interval of Δx. For each value of x in the interval ($x = 2.0$, $x = 2.5$, $x = 3.0$, etc.) the slope of the function differs. However, if we let Δx approach 0, we can obtain an accurate approximation of the slope for a selected value of x. Applying the formula for the instantaneous rate of change, Formula (1.6), to the function $f(x) = x^2/4$, we obtain

$$f'(x) = \operatorname*{limit}_{\Delta x \to 0} \frac{f(x + \Delta x) - f(x)}{\Delta x} = \operatorname*{limit}_{\Delta x \to 0} \frac{\dfrac{(x + \Delta x)^2}{4} - \dfrac{x^2}{4}}{\Delta x}$$

$$f'(x) = \operatorname*{limit}_{\Delta x \to 0} \frac{x^2 + 2x(\Delta x) + (\Delta x)^2 - x^2}{4(\Delta x)} = \operatorname*{limit}_{\Delta x \to 0} \left(\frac{2x(\Delta x)}{4(\Delta x)} + \frac{(\Delta x)^2}{4(\Delta x)}\right)$$

$$f'(x) = \operatorname*{limit}_{\Delta x \to 0} \left(\frac{2x}{4} + \frac{\Delta x}{4}\right).$$

The limiting value of $2x/4$ as Δx approaches 0 is $2x/4$, and the limiting value of $\Delta x/4$ as Δx approaches 0 is 0. Therefore, the derivative of $f(x) = x^2/4$ is

$$f'(x) = \frac{2x}{4}.$$

The slope of $f(x)$ can be determined at any value $x = a$, simply by evaluating the derivative at the value of $x = a$. Remembering that $x = a$ is a value of the independent variable, we represent the value of the derivative of the function at $x = a$ by the symbol $f'(a)$. The slope of the function $f(x) = x^2/4$ at $x = 2$, $x = 3$, and $x = 4$ is $f'(2) = 2(2)/4 = 1$; and similarly $f'(3) = 1.5$ and $f'(4) = 2$.

We can check our formula by selecting an arbitrarily small value for Δx, and determining the *average slope* of the function between $x = a$ and $x = a + \Delta x$. If, for example, $\Delta x = 0.01$ and $x = 2$, the average slope is

$$\text{Slope} = \frac{f(x + \Delta x) - f(x)}{\Delta x} = \frac{(2.01)^2 - (2.0)^2}{4(0.01)}$$

$$\text{Slope} = \frac{4.0401 - 4.0000}{0.04} = 1.0025.$$

This is a close approximation to the instantaneous rate of change of $f'(2) = 1.0$.

The function graphed in Fig. 1.8 is of the form $f(x) = bx^2$. It is possible to apply the formula for the derivative directly to the general form of the

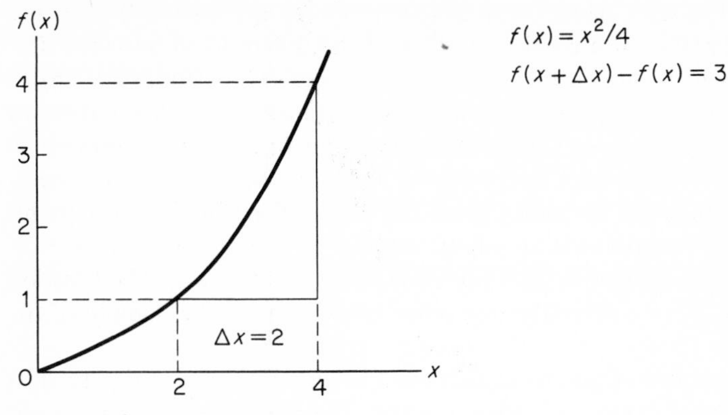

Figure 1.8

function. Once the derivative of the general function $f(x) = bx^2$ has been determined, we can determine the derivative of all functions of this form without resulting to continual reapplication of Formula (1.6). The derivative of $f(x) = bx^2$ is

$$f'(x) = \lim_{\Delta x \to 0} \frac{b(x + \Delta x)^2 - bx^2}{\Delta x}$$

$$f'(x) = \lim_{\Delta x \to 0} \frac{b(x^2 + 2x\,\Delta x + \Delta x^2) - bx^2}{\Delta x}$$

$$f'(x) = \lim_{\Delta x \to 0} \frac{bx^2 + 2bx\,\Delta x + b\,\Delta x^2 - bx^2}{\Delta x}$$

$$f'(x) = \lim_{\Delta x \to 0} \frac{2bx\,\Delta x + b\,\Delta x^2}{\Delta x}$$

$$f'(x) = \lim_{\Delta x \to 0} (2bx + b\,\Delta x)$$

$$f'(x) = 2bx.$$

Using this rule for the derivative of $f(x) = bx^2$, the derivative of $f(x) = x^2/4$ is $f'(x) = 2x/4$. The use of rules for determining derivatives is discussed in the following section.

The concepts of continuity and limit have been necessary for the development of the concept of the derivative. *The derivative of a function gives the slope of the function at a value of the independent variable.* If a function, as shown in Fig. 1.6(c), is discontinuous, it has no slope for those values of the independent variable in the interval of the discontinuity. Consequently, a function must be continuous to have a derivative.

The slope is given by the change in $f(x)$ divided by the change in x. The limit allows us to determine the slope for progressively smaller values of a change in x. The limit of the ratio $\Delta f(x)/\Delta x$ as Δx approaches 0 gives the slope of the function at a particular value of the independent variable. The limit of this ratio is termed the derivative and is given by Formula (1.6).

That a function is continuous is a necessary but not sufficient condition for it to have a derivative. The function shown in Fig. 1.4(b) is continuous at the point $x = a$. It does not, however, have a derivative at this point. Remembering that the derivative gives the slope of a function at a point, observe that the function has an infinite number of possible slopes at $x = a$. Consequently, the derivative does not exist at the point $x = a$. The function does, however, have a derivative for other values of x in the domain of the function.

The mathematical requirement that must be met for a derivative to exist is defined in terms of the derivative formula. The requirement is that

$$\underset{\substack{\Delta x \to 0 \\ \text{from the left}}}{\text{limit}} \left[\frac{f(x+\Delta x) - f(x)}{\Delta x}\right] = \underset{\substack{0 \leftarrow \Delta x \\ \text{from the right}}}{\text{limit}} \left[\frac{f(x+\Delta x) - f(x)}{\Delta x}\right].$$

This requirement is not satisfied by the function in Fig. 1.4(b). The slope for an incremental Δx to the right of a differs from that for an incremental Δx to the left of a.

It should be remembered that the derivative gives the rate of change or slope of the function. To determine the slope of the function at a particular value of the independent variable, the derivative must be evaluated for this value of the independent variable. The following symbols are used for the function, the derivative, and the slope of the function:

$f(x)$ represents the function.
$f'(x)$ represents the derivative of the function.
$f'(a)$ represents the slope of the function
at the value $x = a$ of the independent variable.

Alternative symbols that are commonly used are $y = f(x)$ for the function, dy/dx for the derivative, $dy/dx\ (x = a)$ for the slope of the function at $x = a$. Note that $f'(x)$ and dy/dx are used interchangeably to indicate the derivative. Other commonly used symbols that indicate derivatives are y', f_x, and df/dx.

1.2.1 RULES FOR DETERMINING DERIVATIVES

Rules for determining the derivatives of algebraic functions can now be given.† The formula for the derivative is used to derive the first two rules. All of the remaining rules are given without derivation.

---

†Algebraic functions are an important class of functions. These are defined in Sec. 1.2.3.

Rule 1. The derivative of a constant is 0.

$$f(x) = c, \qquad f'(x) = 0.$$

Derivation: Let $f(x) = c$; then

$$f'(x) = \lim_{\Delta x \to 0} \frac{f(x + \Delta x) - f(x)}{\Delta x} = \lim_{\Delta x \to 0} \frac{c - c}{\Delta x} = 0.$$

Example. $f(x) = 6$; then $f'(x) = 0$.

Example. $f(x) = 3$; then $f'(x) = 0$.

The slope of each of these functions is 0. That is, $f'(a) = 0$ for all values of $x = a$ in the domain. That the slope of the function is 0 for all values of x is easily seen from Fig. 1.9. The function $f(x) = c$ is plotted in this figure. Since the function plots as a horizontal line, the slope of the function is 0 at all values of x.

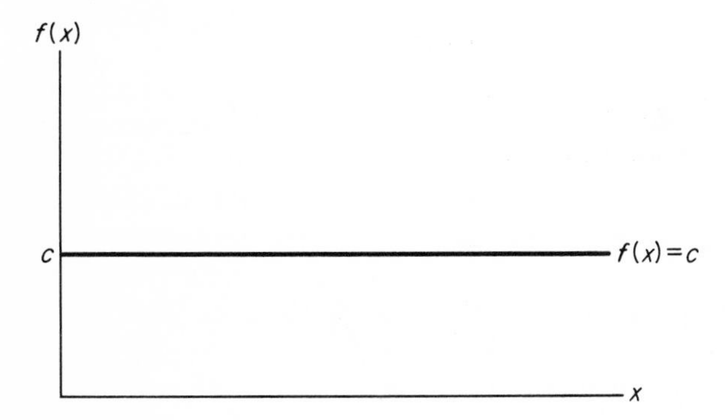

Figure 1.9

Rule 2. The derivative of any variable that is raised to the first power is 1.

$$f(x) = x, \qquad f'(x) = 1.$$

Derivation: Let $f(x) = x$; then

$$f'(x) = \lim_{\Delta x \to 0} \frac{f(x + \Delta x) - f(x)}{\Delta x} = \lim_{\Delta x \to 0} \frac{x + \Delta x - x}{\Delta x}$$

$$f'(x) = \frac{\Delta x}{\Delta x} = 1.$$

The functional relationship $f(x) = x$ is plotted in Fig. 1.10. It can be seen from this figure that the function plots as a straight line passing through the origin with a slope of 1. For any value of x, the slope of the linear function is constant and equals 1. This can also be determined from the derivative of the function. The derivative of $f(x) = x$ from Rule 2 is $f'(x) = 1$. For any value of $x = a$, the derivative of the function has the constant value of 1.

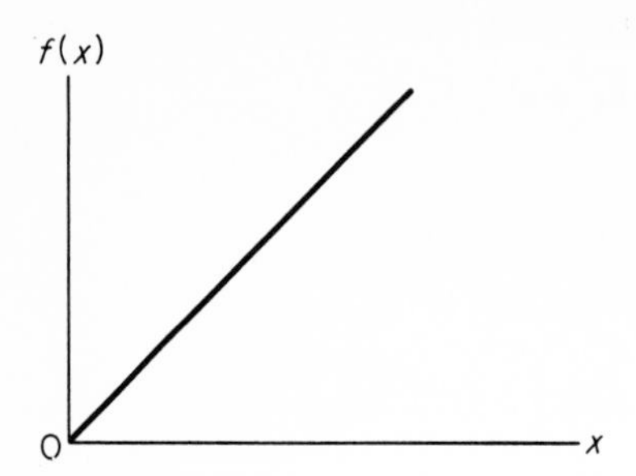

Figure 1.10

Rule 3. The derivative of a variable raised to the constant power n is equal to the power n multiplied by the variable raised to the power $n - 1$.

$$f(x) = x^n, \qquad f'(x) = nx^{n-1}.$$

Example. $f(x) = x^3$; determine the derivative.

$$f'(x) = 3x^2.$$

Example. $y = x^{-3}$; determine the slope at $x = 3$.

$$\frac{dy}{dx} = -3x^{-4} = -\frac{3}{x^4}$$

and

$$\frac{dy}{dx}(x = 3) = -\frac{3}{(3)^4} = -\frac{3}{81}.$$

Example. $f(x) = x^1$; determine the derivative.

$$f'(x) = 1x^0 = 1,$$

which verifies Rule 2.

Example. $f(x) = \sqrt{x} = x^{1/2}$; determine the slope at $x = 4$.

$$f'(x) = \frac{1x^{-1/2}}{2} = \frac{1}{2\sqrt{x}},$$

and

$$f'(4) = \frac{1}{2\sqrt{4}} = \frac{1}{4}.$$

Rule 4. The derivative of a constant times a function equals the constant times the derivative of the function.

$$f(x) = c \cdot g(x), \qquad f'(x) = c \cdot g'(x).$$

Example. $f(x) = 3x^2$; determine the derivative.

$$f'(x) = 3 \cdot 2 \cdot x^1 = 6x.$$

Example. $y = -4x^{-3}$; determine the slope at $x = -2$.

$$\frac{dy}{dx} = -4(-3)x^{-4} = 12x^{-4} = \frac{12}{x^4},$$

and

$$\frac{dy}{dx}(x = -2) = \frac{12}{16} = \frac{3}{4}.$$

Example. $f(x) = c^2x^4$.

$$f'(x) = c^2(4)x^3 = 4c^2x^3,$$

since c is a constant. c^2 is also a constant, and Rule 4 applies.

Rule 5. The derivative of the sum (or difference) of two or more functions equals the sum (or difference) of their respective derivatives.

$$h(x) = f(x) + g(x), \qquad h'(x) = f'(x) + g'(x),$$
$$h(x) = f(x) - g(x), \qquad h'(x) = f'(x) - g'(x).$$

Example. $f(x) = x^2 + 3x + 6$; determine the derivative.

$$f'(x) = 2x + 3.$$

Example. $f(x) = 3x + 4$; determine the slope at $x = 3$.

$$f'(x) = 3$$

and the slope is constant for all values of x.

Example. $f(x) = 2x^2 + 3x - 4$; determine the slope at $x = 4$.

$$f'(x) = 4x + 3,$$

and

$$f'(4) = 4(4) + 3 = 19.$$

Example. $y = 2x^2 - 3x + 10$; determine the value of x such that the slope is 0. For the function to have a slope of 0, the derivative evaluated at $x = a$ must be 0.

$$\frac{dy}{dx} = 4x - 3,$$

Equating the derivative with 0 gives

$$4x - 3 = 0$$
$$x = \tfrac{3}{4}.$$

The derivative is 0 at $x = \frac{3}{4}$. Consequently, the function has a slope of 0 at $x = \frac{3}{4}$.

Rule 6. The derivative of a product of two functions is equal to the derivative of the first function times the second function plus the first function times the derivative of the second function.

$$h(x) = f(x) \cdot g(x), \qquad h'(x) = f'(x) \cdot g(x) + f(x) \cdot g'(x).$$

Example. $h(x) = x^3(3x^2 + 4x)$; determine the derivative.

$$h'(x) = 3x^2(3x^2 + 4x) + x^3(6x + 4).$$

Example. $y = (x^2 + 3)(x + 2)$; determine the derivative.

$$\frac{dy}{dx} = (2x)(x + 2) + (x^2 + 3)(1).$$

Example. $z = u^2(u - 3)$; determine values of u such that the slope of the function is 0.

$$\frac{dz}{du} = 2u(u - 3) + u^2(1) = 3u^2 - 6u.$$

The slope of z is 0 for values of u that make the derivative equal 0. The derivative is equated with 0 to determine the values of u.

$$3u^2 - 6u = 0$$
$$3u(u - 2) = 0.$$

The function z has a slope of 0 for $u = 0$ and for $u = 2$.

Rule 7. The derivative of the quotient of two functions is equal to the denominator times the derivative of the numerator minus the numerator times the derivative of the denominator, all divided by the square of the denominator.

$$h(x) = \frac{f(x)}{g(x)}, \qquad h'(x) = \frac{g(x)f'(x) - f(x)g'(x)}{g(x)^2}.$$

Example. $h(x) = \dfrac{2x^2 + x}{x^3 - 3}$; determine the derivative

$$h'(x) = \frac{(x^3 - 3)(4x + 1) - (2x^2 + x)(3x^2)}{(x^3 - 3)^2}.$$

Example. $h(x) = \dfrac{(2x + 3)}{(x + 2)}$; determine the slope at $x = 2$.

$$h'(x) = \frac{(x + 2)(2) - (2x + 3)(1)}{(x + 2)^2} = \frac{1}{(x + 2)^2},$$

$$h'(2) = \frac{1}{(2 + 2)^2} = \frac{1}{16}.$$

Example. $y = \dfrac{x}{x^2}$; determine the derivative.

$$\frac{dy}{dx} = \frac{x^2(1) - x(2x)}{x^4} = \frac{-x^2}{x^4} = \frac{-1}{x^2}.$$

Note that this checks with the result obtained by reducing the function to $y = x^{-1}$ and using Rule 3.

1.2.2 DERIVATIVES OF COMPOSITE FUNCTIONS

A composite function occurs when one function can be considered as a variable in another function. That is,

$$y = f(g(x)) \tag{1.7}$$

can be decomposed into

$$y = f(u) \quad \text{and} \quad u = g(x).$$

Example. $y = \sqrt{x^2 + 2x + 4}$ can be decomposed into

$$y = \sqrt{u} \quad \text{and} \quad u = x^2 + 2x + 4.$$

Example. $y = (2x + 1)^3$ can be decomposed into

$$y = u^3 \quad \text{and} \quad u = 2x + 1.$$

Example. $y = (x + 2)^3$ can be decomposed into

$$y = u^3 \quad \text{and} \quad u = x + 2.$$

Example. $y = (4 - x^3)^6$ can be decomposed into

$$y = u^6 \quad \text{and} \quad u = 4 - x^3.$$

Derivatives of composite functions of the form

$$y = f(g(x)),$$

where

$$y = f(u) \quad \text{and} \quad u = g(x),$$

can be obtained by first decomposing the function and applying the rule for determining the derivatives of such functions. This rule, given below, is termed the *chain rule.*

Rule 8. If y is a differentiable function of u, $y = f(u)$, and u is a differentiable function of x, $u = g(x)$, then

$$\frac{dy}{dx} = \frac{dy}{du} \cdot \frac{du}{dx} \cdot$$

Example. $y = (3x)^2$; determine $\dfrac{dy}{dx} \cdot$ Let

$$y = u^2 \quad \text{and} \quad u = 3x.$$

$$\frac{dy}{dx} = \frac{dy}{du} \cdot \frac{du}{dx},$$

where

$$\frac{dy}{dx} = 2u; \qquad \frac{du}{dx} = 3.$$

$$\frac{dy}{dx} = (2u)3 = 6u, \quad \text{or} \quad \frac{dy}{dx} = 6(3x) = 18x.$$

Example. $y = (x^2 + 3x + 4)^4$. Let

$$y = u^4 \quad \text{and} \quad u = x^2 + 3x + 4.$$

$$\frac{dy}{dx} = \frac{dy}{du} \cdot \frac{du}{dx}$$

where

$$\frac{dy}{du} = 4u^3; \qquad \frac{du}{dx} = 2x + 3$$

$$\frac{dy}{dx} = 4u^3(2x + 3) = 4(x^2 + 3x + 4)^3(2x + 3).$$

We shall use the chain rule to develop a rule for determining the derivative of composite functions. We have shown that a composite function of the type

$$y = (x^2 + 4x)^3$$

can be decomposed and written as $y = u^3$ and $u = x^2 + 4x$. If y is a differentiable function of u and u is a differentiable function of x, then $y = (x^2 + 4x)^3$ has a derivative of y with respect to x, which is given by the chain rule. Thus,

$$\frac{dy}{dx} = \frac{dy}{du} \cdot \frac{du}{dx} = 3u^2(2x + 4) = 3(x^2 + 4x)^2(2x + 4).$$

Examination of this derivative verifies the following general rule for differentiating composite functions.

Rule 9. The derivative of a function raised to the power n is equal to the power n times the function raised to the power $n - 1$, all multiplied by the derivative of the function.

$$h(x) = [f(x)]^n \qquad h'(x) = n[f(x)]^{n-1} \cdot f'(x).$$

Example. $h(x) = [f(x)]^n = (x^2 + 4x)^3$; determine the derivative.

$$h'(x) = n[f(x)]^{n-1}f'(x) = 3(x^2 + 4x)^2(2x + 4).$$

Example. $y = (x^3 + 2x^2 + 4)^5$; determine the derivative.

$$\frac{dy}{dx} = 5(x^3 + 2x^2 + 4)^4(3x^2 + 4x).$$

Example. $y = (x^2 - 4x + 4)^3$; determine the slope at $x = \frac{1}{2}$.

$$\frac{dy}{dx} = 3(x^2 - 4x + 4)^2(2x - 4)$$

$$\frac{dy}{dx}(x = \tfrac{1}{2}) = 3(\tfrac{1}{4} - 2 + 4)^2(1 - 4) = 3(2.25)^2(-3)$$

$$= -45.56.$$

Example. $y = \sqrt{(2x^2 - x + 4)}$; determine the slope at $x = 2$.

$$y = (2x^2 - x + 4)^{1/2}$$

$$\frac{dy}{dx} = \frac{1}{2}(2x^2 - x + 4)^{-1/2}(4x - 1) = \frac{4x - 1}{2\sqrt{2x^2 - x + 4}}$$

$$\frac{dy}{dx}(x = 2) = \frac{8 - 1}{2\sqrt{8 - 2 + 4}} = \frac{7}{2\sqrt{10}} = 1.11.$$

1.2.3 COMBINING THE RULES

The rules for determining derivatives discussed in this chapter apply to polynomials of the form

$$y = a_n x^n + a_{n-1}x^{n-1} + a_{n-2}x^{n-2} + \cdots + a_0, \tag{1.8}$$

where $a_n, a_{n-1}, \ldots, a_0$ represent constants and x represents the independent variable. Functions of this form are termed algebraic functions. The derivative of the function is determined by combining the appropriate rules for derivatives given in this chapter and applying these rules to the function.

Example. $y = x^4 + 4x^3 - 2x^2 + 10x - 25$

$$\frac{dy}{dx} = 4x^3 + 12x^2 - 4x + 10.$$

Example. $y = 3x^{3/2} - x^{1/2} + 2x^{-1/2} + 10$

$$\frac{dy}{dx} = \frac{9}{2}x^{1/2} - \frac{1}{2}x^{-1/2} - x^{-3/2}.$$

Algebraic functions also are of the form

$$y = a_n u_n^n + a_{n-1}u_{n-1}^{n-1} + \cdots + a_0, \tag{1.9}$$

where $a_n, a_{n-1}, \ldots, a_0$ again represent constants. $u_n, u_{n-1}, \ldots$ are functions of x. Note that u_n and u_{n-1} need not have the same functional form. The derivative of y with respect to x is determined by applying the appropriate rules for derivatives.

Example. $y = 6(x^2 + 2x + 10)^4 + 100x^2 + 500$

$$\frac{dy}{dx} = 24(x^2 + 2x + 10)^3(2x + 2) + 200x.$$

Example.
$$y = 4(2x + 2)^2 + 6(2x + 2) + 2$$
$$\frac{dy}{dx} = 8(2x + 2)(2) + 12$$
$$\frac{dy}{dx} = 16(2x + 2) + 12.$$

Example.
$$y = 3(x^2 + 2x)^3 + 4(x + 1)^2$$
$$\frac{dy}{dx} = 9(x^2 + 2x)^2(2x + 2) + 8(x + 1).$$

Still another type of algebraic function is of the form

$$y = a_n f_n(x)g_n(x) + a_{n-1}f_{n-1}(x)g_{n-1}(x) + \cdots + a_0, \tag{1.10}$$

where a_n, a_{n-1}, . . . , a_0 again represent constants, and $f_n(x)$ and $g_n(x)$ represent functions of x. The derivative is determined by combining the appropriate rules as illustrated by the following examples.

Example.
$$y = 3x^2(x^2 + 2x)^3 + 2x(x + 5)^2$$
$$\frac{dy}{dx} = 6x(x^2 + 2x)^3 + 3x^2[3(x^2 + 2x)^2(2x + 1)]$$
$$+ 2(x + 5)^2 + 2x[2(x + 5)(1)].$$

Example.
$$y = (x + 3)^2(x^2 + 2x)^3$$
$$\frac{dy}{dx} = 2(x + 3)(x^2 + 2x)^3 + (x + 3)^2[3(x^2 + 2x)^2(2x + 2)].$$

These examples, it should be noted, required the use of the derivative rules for a constant times a function, the sum of two functions, the product of two functions, and the composite function. In fact, the only rule discussed in this chapter that was not applied in one of the preceding two examples was that for the quotient of two functions. We shall next consider algebraic functions whose derivative requires application of the quotient rule.

The function

$$y = \frac{f(x)}{g(x)},$$

where $f(x)$ and $g(x)$ are polynomials similar to those defined by (1.8), (1.9), or (1.10), is an algebraic function. Again, the derivative of this function is determined by application of the appropriate rules of derivatives. This type of function is illustrated by the following examples.

Example. $y = \dfrac{(x^2 + 3x)^3}{(x^3 + 4x^2)^2}$; determine the derivative. To determine the derivative of this function requires application of the quotient rule, the composite function rule, the power rule, and the rule for a constant times a function. The derivative of the function is

$$\frac{dy}{dx} = \frac{(x^3 + 4x^2)^2 3(x^2 + 3x)^2(2x + 3) - (x^2 + 3x)^3 2(x^3 + 4x^2)(3x^2 + 8x)}{(x^3 + 4x^2)^4}.$$

Example. $y = \frac{(2x^3 + 4x^2)^2}{x^2 + 3}$; determine the derivative. We again apply the rules illustrated in this chapter. The derivative is

$$\frac{dy}{dx} = \frac{(x^2 + 3)2(2x^3 + 4x^2)(6x^2 + 8x) - (2x^3 + 4x^2)^2(2x)}{(x^2 + 3)^2}.$$

These examples illustrate determining the derivatives of algebraic functions. The procedure is to apply the appropriate combination of rules as required by the function. This, of course, requires the ability to recognize the functional form of the function and to be able to select the appropriate rule. This skill normally comes with a reasonable amount of practice.

1.2.4 INVERSE FUNCTIONS

The algebraic functions discussed in this chapter have been of the form $y = f(x)$, where y is the dependent variable and x is the independent variable. If, given the function $y = f(x)$, we rewrite the function with x as the dependent variable and y as the independent variable, the resulting function $x = g(y)$ is termed the inverse of $y = f(x)$.

As an example of inverse functions, consider the function $y = 4x + 10$. The inverse of this function is $x = 0.25y - 2.5$. The inverse was found by solving the function for x in terms of y; that is, $x = g(y)$.

The functions considered in this chapter have been *explicit* functions. Functions such as $y = 4x - 3$, in which the dependent variable is clearly designated as y and the independent variable as x, are termed explicit functions. A function can also be defined *implicitly*. As an example, the equation $5x - 2y = 14$ implicitly defines x in terms of y. If y is allowed to assume a certain value, then x is implicitly defined by the equation. Similarly, y is an implicit function of x in that if x is permitted to take on certain values, the value of y is implicitly established by the equation. If we are given the equation $h(x, y) = 0$, then $y = f(x)$ and $x = g(y)$ are inverse functions. The relationship between implicit functions and inverse functions is illustrated by the following examples.

Example. Consider the implicit function $3x + 4y - 6 = 0$. Determine $y = f(x)$ and $x = g(y)$ for this function. The explicit function of y in terms of x is

$$y = -0.75x + 1.5$$

and the explicit function of x in terms of y is

$$x = -1.33y + 2.$$

$y = f(x)$ and $x = g(y)$ are inverse functions.

Example. For the implicit function $x + 2y - 10 = 0$, determine the explicit functions $y = f(x)$ and $x = g(y)$. The explicit function $y = f(x)$ is

$$y = 5 - 0.5x$$

and the explicit function $x = g(y)$ is

$$x = 10 - 2y.$$

Example. Determine the inverse of the explicit function $y = x^2$. This is an example of a function that has more than one inverse function. By expressing x in terms of y, we find that

$$x = +\sqrt{y} \quad \text{and} \quad x = -\sqrt{y}$$

are both inverse functions of $y = x^2$.

It is not necessary to determine the inverse of a function in order to determine the derivative of that inverse. The derivatives of inverse functions can be determined by application of the following rule.

Rule 10. If $x = g(y)$ is the inverse function of $y = f(x)$, the derivative of the inverse function, if it exists, is given by the reciprocal of the derivative of $y = f(x)$. Thus

$$\frac{dx}{dy} = \frac{1}{dy/dx}, \quad \text{provided that } \frac{dy}{dx} \neq 0.$$

The rule for determining derivatives of inverse functions is illustrated by the following examples.

Example. For the implicit function $y - 3x - 4 = 0$, determine $y = f(x)$, dy/dx, $x = g(y)$, and dx/dy. The explicit function $y = f(x)$ is

$$y = 3x + 4$$

and has the derivative

$$\frac{dy}{dx} = 3.$$

The explicit function $x = g(y)$ is

$$x = \frac{y}{3} - \frac{4}{3}$$

and has the derivative

$$\frac{dx}{dy} = \frac{1}{3}.$$

The derivative dx/dy can also be determined from the rule for derivatives of inverse functions. Thus

$$\frac{dx}{dy} = \frac{1}{dy/dx} = \frac{1}{3}.$$

Example. For $y = 2x^2 + 3x + 6$, determine dy/dx and dx/dy.

$$\frac{dy}{dx} = 4x + 3.$$

From the rule for derivatives of inverse functions

$$\frac{dx}{dy} = \frac{1}{dy/dx} = \frac{1}{4x+3}, \quad \text{provided that} \quad x \neq \frac{-3}{4}.$$

Example. For the quadratic function $y = x^2 + 2x + 6$, determine both dy/dx and dx/dy. The derivative of y with respect to the independent variable x is

$$\frac{dy}{dx} = 2x + 2.$$

The derivative of x with respect to the independent variable y is

$$\frac{dx}{dy} = \frac{1}{dy/dx} = \frac{1}{2x+2}, \quad \text{provided that} \quad x \neq -1.$$

problems

1. Plot the following functions and specify the slope of each function.
   - a. $f(x) = 4$
   - b. $f(x) = x + 3$
   - c. $f(x) = 2x + 3$
   - d. $f(x) = 4 - 3x$
   - e. $f(x) = -0.5x + 2$
   - f. $f(x) = -x$
   - g. $f(x) = -3x - 4$
   - h. $f(x) = \frac{x}{3} - 2$
2. Plot the following functions and specify the slope of each function in the interval $x = 2.00$ to $x = 2.01$.
   - a. $f(x) = x^2$
   - b. $f(x) = x^2 + 2x$
   - c. $f(x) = -x^2 + 3x$
   - d. $f(x) = x^2 - 3x + 5$
   - e. $f(x) = x^2 - x + 6$
   - f. $f(x) = x^3$
   - g. $f(x) = x^3 - 2x^2 + 3x + 4$
   - h. $f(x) = -x^3 + 3x^2 + 2x - 8$
3. Determine the derivative of the following functions.
   - a. $f(x) = x^2$
   - b. $f(x) = 2x^2$
   - c. $f(x) = x^5$
   - d. $f(x) = x^{21}$
   - e. $f(x) = x^{-4}$
   - f. $f(x) = x^{-1/2}$
   - g. $f(x) = x^{1/2}$
   - h. $f(x) = x^{0.35}$
   - i. $f(x) = x^{1/4}$
   - j. $f(x) = x^{-1/4}$

k. $f(x) = 1/x^2$
l. $f(x) = 1/x^4$
m. $f(x) = x^n$
n. $f(x) = x^a$
o. $f(x) = a^b$ ≃ 0
p. $f(x) = b^a$ = 0
q. $f(x) = \sqrt{x}$
r. $f(x) = 3\sqrt{x}$

4. Determine the derivative of the following functions.
   a. $f(x) = 4x$
   b. $f(x) = x/2$
   c. $f(x) = -3x$
   d. $f(x) = 2x^{1/2}$
   e. $f(x) = x^4/4$
   f. $f(x) = 3x^2/2$
   g. $f(x) = 2x^{-3}/5$
   h. $f(x) = 6x^{-1}/3$
   i. $f(x) = ax^b$
   j. $f(x) = -ax^{-b}$
   k. $f(x) = 3/x^2$
   l. $f(x) = 4/x^7$
   m. $f(x) = k/x^{1/2}$
   n. $f(x) = ab/x^4$

5. Determine the derivative of the following functions.
   a. $f(x) = 3x^2 + 2x + 5$
   b. $f(x) = 4x^3 + 3x^2 - 10$
   c. $f(x) = 25x^3 + 16x^2 + 6x + 7$
   d. $f(x) = -6x^{-2} - 5x^{-3}$
   e. $f(x) = 3x^{1/4} + 2x^{1/2}$
   f. $f(x) = 1/x^2 + 3/x^3$
   g. $f(x) = ax^4 + bx^3 + cx^2 + dx + k$
   h. $f(x) = 5x^5 + 6x^4 - 10x^3 + 16x^2 + 25x + 100$
6. Determine the derivative of the following functions.
   a. $f(x) = 2x^2(x + 3)$
   b. $f(x) = 3x(x^3 + 2x + 4)$
   c. $f(x) = (x + 6)(2x + 3)$
   d. $f(x) = (x^2 + 2x)(x^3 + 3x + 5)$
   e. $f(x) = (x^5 - 6)(x^4 - 5)$
   f. $f(x) = (x^{1/2} + x)(x^{1/4} + 2x)$
   g. $f(x) = (x^3 - 2x^2 + 4x + 10)(x^2 + 4x + 17)$
   h. $f(x) = (x^{-2} + 4x^{-3})(x^3 + 6x^2 + 12x + 24)$
7. Determine the derivative of the following functions.
   a. $f(x) = \dfrac{x - 2}{x + 3}$
   b. $f(x) = \dfrac{2x + 4}{x + 6}$
   c. $f(x) = \dfrac{2x^2 - 5x}{x^2 + 2}$
   d. $f(x) = \dfrac{2x^3 + 3x^2 + 2x}{x^2 + 2x + 1}$
   e. $f(x) = \dfrac{x^{-3} + x^{-2}}{x^4 + 3x^2}$
   f. $f(x) = \dfrac{x^{-1/2} + x^{-1/2}}{x^{-1/4} + x^{-1/2}}$
   g. $f(x) = \dfrac{(x^2 + 4x)(x^3 + 3x^2)}{x^2 + 2x}$
   h. $f(x) = \dfrac{2x^2(x^4 + 4x)}{x + 3}$
   i. $f(x) = \dfrac{(x^2 + 5x)(3x + 4)x^2}{(x + 3)(x^2 + 3x)}$
8. Determine the derivative of the following functions.
   a. $f(x) = (x^2 + 2x)^3$
   b. $f(x) = (x^3 + 3x^2 + 2x + 5)^4$
   c. $f(x) = (2x + 3)^{1/2}$
   d. $f(x) = (x^2 + 4x)^{-5}$
   e. $f(x) = (x^4 - 4x^2)^3$
   f. $f(x) = x^2(4x^3 + 3x)^2$
   g. $f(x) = (x^2 + 3)^3(x + 4)^2$
   h. $f(x) = (x + 3)^2(x^2 - 3)^{-2}$
   i. $f(x) = (x + 3)^{1/2}(x + 3)^{-1/2}$
   j. $f(x) = (6x^3 + 3x)^{-3}(x^2 + 4x + 7)$

9. Find the slope of the function at the specified value of the independent variable.

   a. $f(x) = x^2 + 2x$, $f'(3) =$
   b. $f(x) = 3x^2 + 2x - 5$, $f'(4) =$
   c. $f(x) = 2x^2 - 3x + 6$, $f'(1) =$
   d. $f(x) = (x + 3)^3$, $f'(5) =$
   e. $f(x) = (2x^2 + 4x)^{1/2}$, $f'(2) =$
   f. $f(x) = x^3 + 2x^2 + 6x$, $f'(-2) =$
   g. $f(x) = \sqrt{3x + 4}$, $f'(-1) =$
   h. $y = 2x^3 + 3x$, $\frac{dy}{dx}(x = 3) =$
   i. $y = (x^2 + 5)^2$, $\frac{dy}{dx}(x = 2) =$
   j. $y = (x^2 + 2x)^{-2}$, $\frac{dy}{dx}(x = 1) =$

10. Find the value of the independent variable for which the slope of the function is 0.

   a. $f(x) = x^2 - 3x + 6$
   b. $f(x) = x^2 + 4x + 8$
   c. $f(x) = x^2 - 5x + 10$
   d. $f(x) = x^2 - 2x + 4$
   e. $f(x) = \frac{x^3}{3} + \frac{3x^2}{2}$
   f. $f(x) = x^3 - 2x^2$
   g. $f(x) = \frac{x^3}{3} + \frac{x^2}{2} - 12x$
   h. $f(x) = \frac{x^3}{3} - \frac{5x^2}{2} + 6x + 10$

11. Establish linear functions through the following data points $(x, f(x))$.

   a. (3, 4) and (5, 2)
   b. (5, 8) and (−4, 7)
   c. (9, 4) and (4, −1)
   d. (−3, −5) and (5, 8)
   e. (8, 2) and (5, 5)
   f. (3, 9) and (0, −6)
   g. (4, −3) and (−5, 6)
   h. (5, −6) and (7, −2)
   i. (8, 4) and (4, −3)
   j. (7, 3) and (3, −6)
   k. (−4, −5) and (4, 5)
   l. (9, 0) and (0, 9)
   m. (3, 3) with $b = \frac{1}{2}$
   n. (−3, 4) with $b = -\frac{3}{4}$
   o. (−5, 7) with $b = 0.75$
   p. $(\frac{1}{4}, \frac{2}{5})$ with $b = \frac{5}{6}$
   q. $(\frac{3}{4}, \frac{6}{5})$ with $b = -\frac{3}{2}$
   r. $(4, \frac{1}{4})$ with $b = -3$
   s. $(\frac{4}{3}, 70)$ with $b = \frac{3}{4}$
   t. (1.5, 2.4) with $b = 0.67$
   u. (0.8, 0.5) with $b = 0.5$
   v. (0.4, 2.5) with $b = -0.75$
   w. (3, −4) with $b = 5$
   x. (4, 4.5) with $b = \frac{3}{4}$
   y. (0.75, 0.25) with $b = 0.4$
   z. (3.5, 4.5) with $b = 0.60$

12. Establish linear functions for the following problems.

   a. Costs are \$3000 when output is 0 and \$4000 when output is 1500 units.
   b. Fixed costs are \$75,000, and variable costs are \$3.50.

c. Costs are \$5000 for 150 units and \$6000 for 250 units.

d. If 1000 units can be sold at a price of \$2.50 and 2000 can be sold at a price of \$1.75, determine price as a function of quantity.

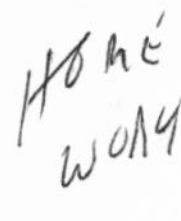

e. If 1500 units can be sold at a price of \$10 and 2500 units can be sold at a price of \$9.00, determine price as a function of quantity.

f. A firm has 250 employees and expects to add to the work force by 10 employees per year. Determine the number of employees as a function of the year.

g. An individual now earning \$10,000 per year expects raises of \$500 per year for the next five years.

h. A dealer advertises that a car worth \$3000 will be reduced by \$100 per month until sold.

2

optimization using calculus

2.1 optimum values of functions

In this chapter we develop a technique based upon differential calculus for determining the optimum value of a function. For example, suppose that profit and quantity are related as shown in Fig. 2.1. Our objective is to determine the quantity so that profits are a maximum.

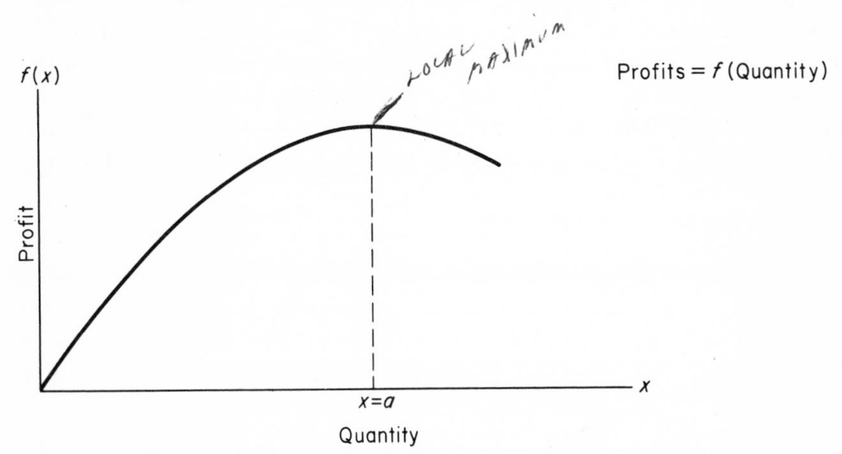

Figure 2.1

Similarly, we might want to minimize the cost of purchasing an object. When both storage costs and ordering costs are taken into account, the cost of purchasing and storing the inventory varies with the quantity ordered as shown in Fig. 2.2. Our objective is to determine the quantity so that cost is

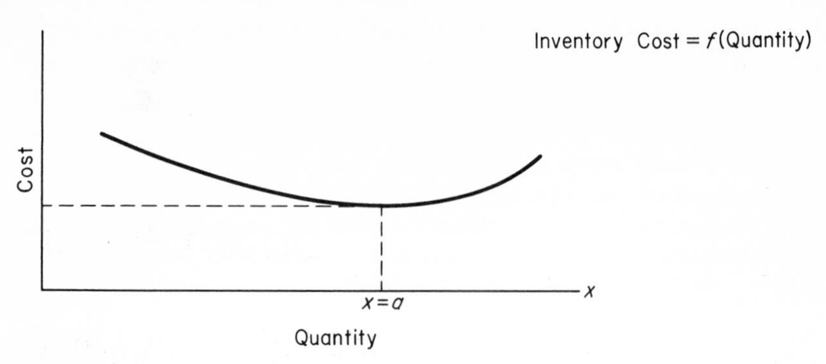

Figure 2.2

a minimum. It is customary to use value of the function and value of the dependent variable interchangeably.

The optimum value of a function can be either a maximum or a minimum. A function is said to reach a local maximum at the value $x = a$ of the independent variable if the value of the dependent variable at $x = a$ is greater than the value of the dependent variable at any adjacent point. Similarly, a function reaches a local minimum at the point $x = a$ if the value of the dependent variable at that point is less than the dependent variable at adjacent points.

As an example of local optimum, consider the function graphed in Fig. 2.3. The function evaluated at $x = b$ is a local maximum since the value of the function is greater at $x = b$ than at any adjacent points, shown as $b + \Delta x$ and $b - \Delta x$. Similarly, the function evaluated at $x = c$ is a local minimum. The quantity Δx is an arbitrarily small increment of the analyst's choosing.

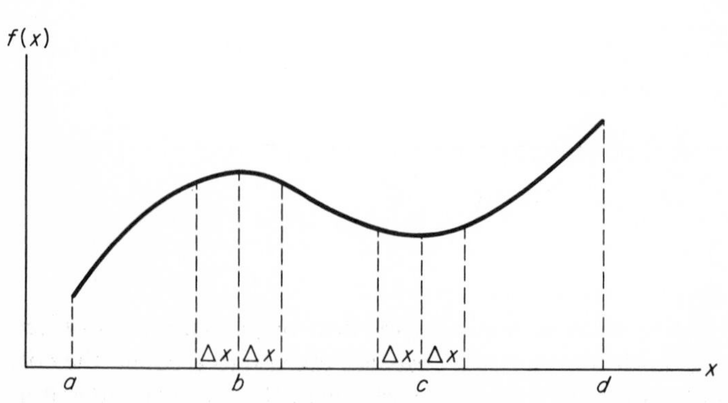

Figure 2.3

What about the value of the dependent variable at $x = a$ and $x = d$? Clearly, the value of the dependent variable is less at $x = a$ than at $x = c$. Also, the value of the dependent variable is greater at $x = d$ than at $x = b$. The function evaluated at $x = a$ is thus termed the *absolute minimum*, and the function evaluated at $x = d$ is termed the *absolute maximum*. An absolute minimum of the function occurs for that value of $x = a$ for which the function evaluated at $x = a$ is smaller than for any other value of x in the domain. Similarly, an absolute maximum of the function occurs for that value of $x = d$ for which the function evaluated at $x = d$ is larger than for any other value of x in the domain.

Example. Specify all maximum and minimum values of $f(x)$ for the following function.

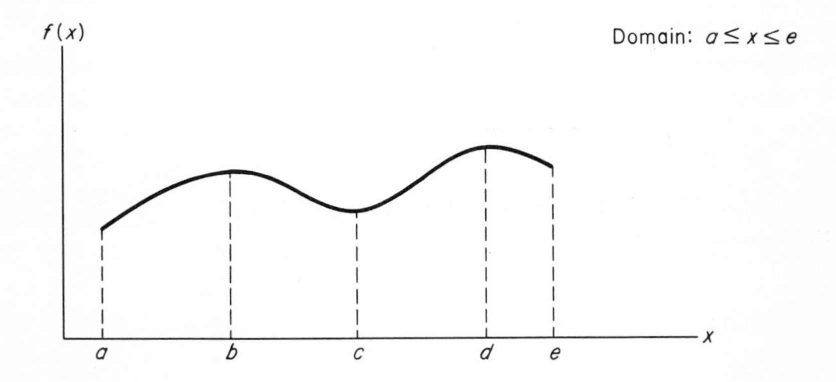

It can be seen that $f(a)$ = absolute minimum, $f(b)$ = local maximum, $f(c)$ = local minimum, and $f(d)$ = absolute maximum.

In the preceding example, the domain of the function is $a \leq x \leq e$. When the end points of the domain, such as a and e, are included as possible values of x, these values of x must be evaluated to determine if the dependent variable is an absolute maximum or minimum. When the end points are included in the domain, the interval between the end points is termed *closed*. If the end points are not included, the interval is said to be *open*. If one end point is included and the other is not, the interval is said to be mixed. A closed interval is written as $a \leq x \leq e$, the open interval is $a < x < e$, and a mixed interval is $a \leq x < e$ or $a < x \leq e$.

End points are not considered as candidates for *local* maxima or minima. The reason for excluding end points is that the function is not defined at both adjacent points. For instance, in the preceding example the function is not defined for values of x larger than $x = e$ or smaller than $x = a$. End points are, however, candidates for *absolute* maxima and minima and must be evaluated. The examples that follow illustrate the concept of optimum values of functions in both open and closed domains of the function.

Example. Find the maximum and minimum value of $f(x)$ with the domain $a \leq x \leq b$.

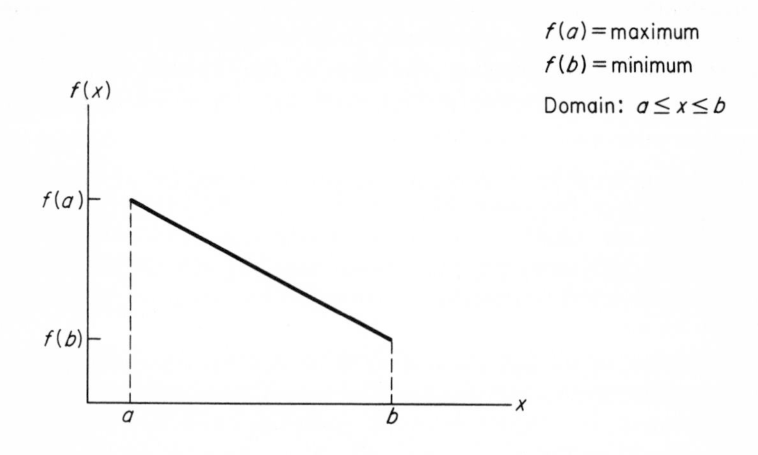

Example. Determine the maximum and minimum values of $f(x)$ with the domain $a < x < d$.

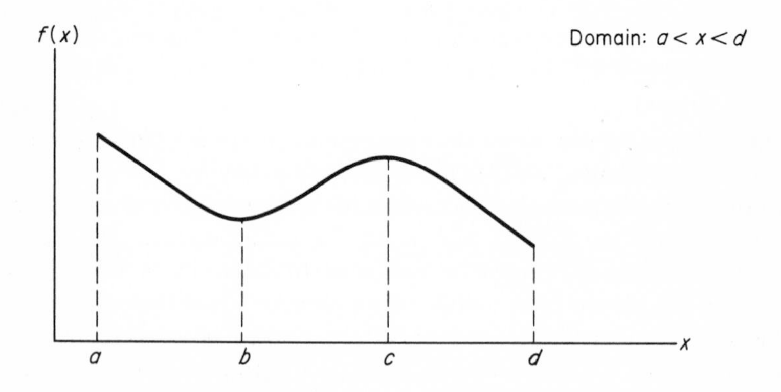

By convention, end points are not considered in determining absolute maxima and minima if the domain of the function is an open interval. Therefore, $f(b)$ is the local minimum and $f(c)$ is the local maximum. We do not specify absolute maxima or minima for functions for which the domain of the function is an open interval.

The preceding example illustrates the case in which the end points of the domain of the function are not considered in determining optimum values of the function. It is quite common in business situations for the analyst to be interested only in local maximum or minimum values of the function. In such cases the end points of the domain are not specified. In this text, we adopt the convention that a function cannot be said to achieve an *absolute* maximum or minimum unless the end points of the independent variable are given and the function is evaluated at these end points. If the end points of the domain are not specified, the function can reach local maxima or minima but not absolute maxima or minima.

If the end points are of no consequence in a problem, the domain of the function is often not specified. For functions for which the domain is not specified, we shall assume that the domain of the function is the open interval $-\infty < x < \infty$. Since the interval is open, we can determine values of the independent variable x for which the function reaches a local optimum. We follow the convention stated above, however, and do not specify an absolute maximum or minimum of a function for which the domain is not explicitly stated.

The domain of the independent variable is often determined from the structure of the problem. For instance, if the independent variable is manufacturing output, the domain of output would be limited on the upper side by the output capacity of the plant. The lower limit of the domain could possibly be that level of output such that the production costs are covered. An example of this type of problem follows.

Example. The Far Western Manufacturing Company is engaged in the business of manufacturing souvenirs that are sold in various shops throughout the western states. The company has found that the volume in standard units of sales is highly dependent upon price; the lower the price the higher the volume. They have also found that if the price is too low, profits tend to fall in spite of an increased volume of sales. Assume that we have been assigned the task of recommending a procedure for establishing the proper volume-profit relationship.

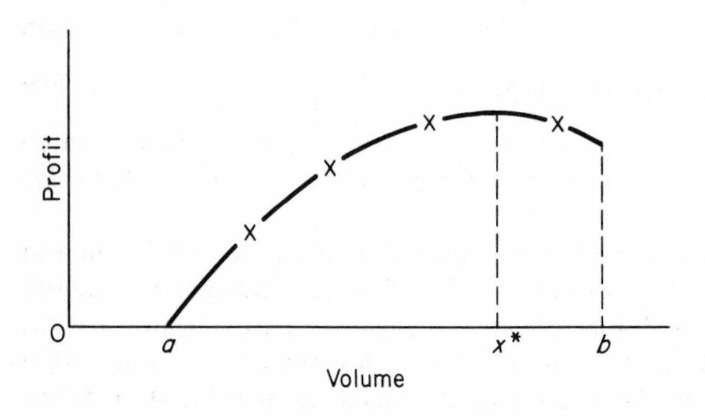

Profit = f(Volume)

Domain: $a \leq x \leq b$

An explanation of the price-profit-volume relationship is presented in Chapter 3. At this point, however, our understanding of the relationship between these variables is sufficient to recommend a study of the historical relationship between changes in profit and volume. Assume that, based upon this study, the profit-volume data plotted on the accompanying graph is obtained. By drawing a curve through these data points, we obtain profit as a function of volume. The volume that leads to maximum profit is indicated by x^*. The price that must be charged to obtain this volume can be determined from company records. The curve is a quadratic function. If data points were given in this example, the parameters of the quadratic function could be determined by the method explained in Appendix A.

2.2 interpretation of derivatives

The derivative of a function gives the slope of the function for alternative values of the independent variable. The slope of the function at a particular value of $x = a$ is determined by evaluating the derivative function at the particular value of $x = a$. Expressed in mathematical notation, the slope is given by $f'(a)$ or $\frac{dy}{dx}(x = a)$.

In understanding the derivative, it is helpful to examine graphs of several functions. First consider the quadratic function

$$y = x^2 - 6x + 8 \qquad \text{for } 0 \leq x \leq 6.$$

The graph of this function is shown in Fig. 2.4(a). This quadratic function plots with the minimum value of y occurring at $x = 3$. The slope of the function is negative for values of x less than 3 and positive for values of x greater than 3. The slope of the function is 0 for $x = 3$.

The derivative of the function is $\frac{dy}{dx} = 2x - 6$. This derivative function gives the slope of the original function, $y = x^2 - 6x + 8$, for all values in the domain of the function. For example, the slope of the original function for $x = 0$ is -6. Similarly, the slope for $x = 1$ is $\frac{dy}{dx}(x = 1) = -4$. The slope of the original function for any value of x in the domain of the function can be determined simply by evaluating the derivative function for that value of x.

From Fig. 2.4(a) it is apparent that the function reaches a minimum at $x = 3$. This can also be seen from the plot of the derivative in Fig. 2.4(b). For values of x less than 3, the derivative is negative. This indicates that the slope of the original function is negative for x less than 3. For values of x greater than 3, the derivative is positive. This indicates that the slope of the original function is positive for x greater than 3. From Fig. 2.4(a), it can be

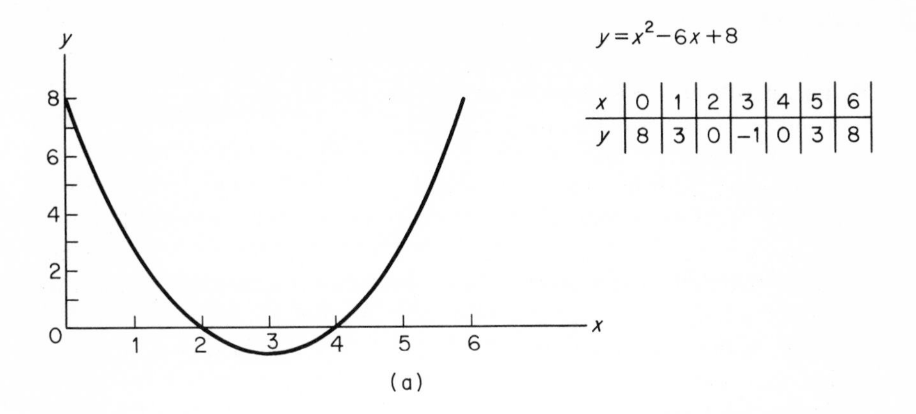

$y = x^2 - 6x + 8$

x	0	1	2	3	4	5	6
y	8	3	0	−1	0	3	8

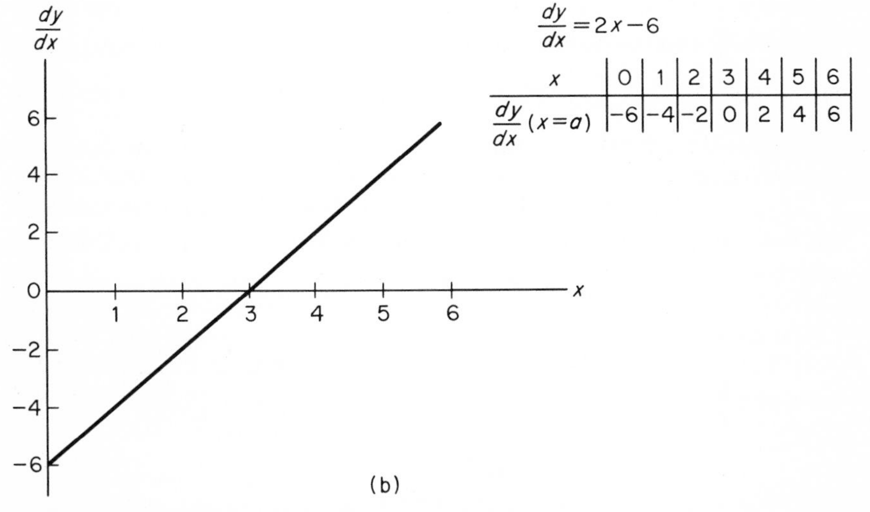

$\frac{dy}{dx} = 2x - 6$

x	0	1	2	3	4	5	6
$\frac{dy}{dx}(x=a)$	−6	−4	−2	0	2	4	6

Figure 2.4

seen that a function whose slope goes from negative to positive reaches a minimum value. This minimum occurs at that value of x for which the slope of the original function is 0. In our example, the slope of the function $y = x^2 - 6x + 8$ is 0 at $x = 3$.

Values of the independent variable for which the slope of the function equals 0 are termed *critical points*. In the last example, $x = 3$ is a critical point. These values of x can be determined from the derivative. To determine the critical point we equate the derivative to 0 and solve for x. For our example

$$\frac{dy}{dx} = 2x - 6 = 0.$$

Therefore,

$$x = 3.$$

Critical points are designated by an asterisk, i.e., $x^* = 3$.

A critical point is a necessary requirement for a local optimum. It does not follow, however, that all critical points result in local optima. Certain critical points result in *points of inflection* of a function. Since critical points can result in points of inflection rather than local optima, a critical point is a necessary but not sufficient condition for a local optimum. It is necessary in that a function cannot reach a local optimum for a value of x unless the slope of the function is 0 for that value of x. A critical point is not sufficient, however, in that critical points can result in points of inflection of a function rather than local optima. A critical point is thus a candidate for a local optimum. Each critical point must be investigated to determine if the critical point is a local optimum or a point of inflection. A critical point that leads to a point of inflection of a function is given by the following example.

Example. Investigate the function $y = x^3$ for local optima. The derivative of the function is $\frac{dy}{dx} = 3x^2$. Equating the derivative function with 0 and solving for the critical point (that is, the value of x such that the derivative function equals 0) gives the critical point $x^* = 0$. From a plot of the function, it can be seen that the critical point $x^* = 0$ is not a local optimum. Rather, the function reaches a point of inflection at this critical point. Points of inflection are defined in Sec. 2.2.4 of this chapter.

x	−3	−2	−1	0	1	2	3
y	−27	−8	−1	0	1	8	27

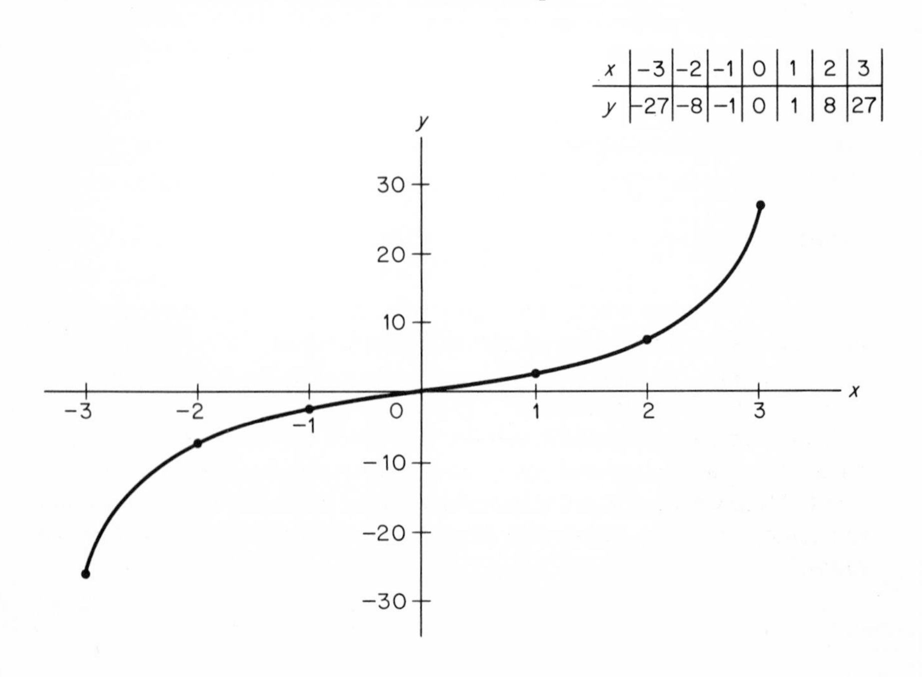

We can summarize the requirements for a local optimum.† A function reaches a local optimum at the critical point $x^* = a$ when $f'(a) = 0$, provided that one of the following conditions is met: (1) The function is a local maximum if $f'(a - \Delta x) > 0$ and $f'(a + \Delta x) < 0$. (2) The function is a local minimum if $f'(a - \Delta x) < 0$ and $f'(a + \Delta x) > 0$. The function reaches an inflection point rather than a local optimum at the critical point $x = a$ if $f'(a - \Delta x)$ and $f'(a + \Delta x)$ are both greater than 0 or are both less than 0.

The relationship between a function and its derivative is further illustrated by the cubic function

$$y = \tfrac{1}{3}x^3 - 4x^2 + 12x + 5.$$

The function is plotted in Fig. 2.5(a), and the derivative function is plotted in Fig. 2.5(b). The function reaches a local maximum at $x = 2$ and a local minimum at $x = 6$. For values of x less than 2, the derivative is positive. This indicates that the function has a positive slope for values of x less than 2. For x greater than 2 but less than 6, the derivative is negative, thus indicating that the slope of the function is negative. Since the slope of the function goes from positive to negative and passes through 0 at $x = 2$, we conclude from the derivative that the function achieves a local maximum at $x = 2$. This conclusion is, of course, verified by inspection of the plot of the function in Fig. 2.5(a).

For values of x between $x = 2$ and $x = 6$, the derivative is negative. For x greater than 6 the derivative is positive. A function whose slope changes from negative to positive necessarily reaches a local minimum. The minimum occurs when the derivative is 0. Thus, from inspection of the derivative at various values of x, we conclude that the function reaches a local minimum at $x = 6$. This conclusion is verified by inspection of the plot of the function.

The values of x for which the function has a slope of 0 can be determined by equating the derivative with 0 and solving the resulting equation for x. In our example,

$$\frac{dy}{dx} = x^2 - 8x + 12 = 0,$$

which can be factored to give

$$(x - 2)(x - 6) = 0.$$

For the left side to equal 0, either $x = 2$ or $x = 6$. The function has a slope of 0 at both $x^* = 2$ and $x^* = 6$.

To determine if the critical point $x^* = 2$ is a local optimum, the derivative can be evaluated at $2 - \Delta x$ and at $2 + \Delta x$. If the slope is positive for

†The use of the second derivative to establish maxima and minima is discussed in Sec. 2.2.3.

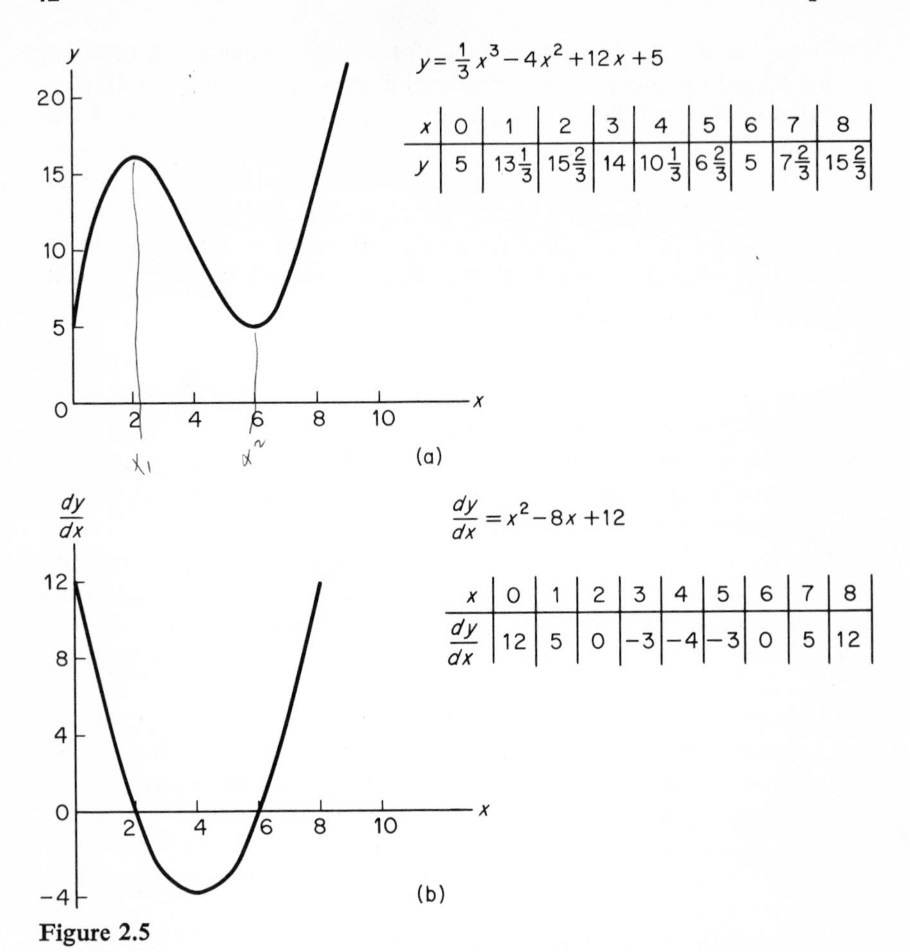

x	0	1	2	3	4	5	6	7	8
y	5	$13\frac{1}{3}$	$15\frac{2}{3}$	14	$10\frac{1}{3}$	$6\frac{2}{3}$	5	$7\frac{2}{3}$	$15\frac{2}{3}$

x	0	1	2	3	4	5	6	7	8
$\frac{dy}{dx}$	12	5	0	−3	−4	−3	0	5	12

Figure 2.5

$2 - \Delta x$ and negative for $2 + \Delta x$, the function has reached a local maximum at the critical point $x^* = 2$. If, however, the derivative is negative at $2 - \Delta x$ and positive at $2 + \Delta x$, the function has reached a minimum at $x^* = 2$. The same procedure can be applied for $x = 6$ to determine if the function is a maximum, minimum, or point of inflection at the critical point $x = 6$.

Example. Find critical points for the following function and determine if the function when valued at the critical points is a maximum, minimum, or point of inflection.

$$y = \frac{x^3}{3} + 4x^2 - 20x + 10,$$

$$\frac{dy}{dx} = x^2 + 8x - 20 = 0,$$

$$(x - 2)(x + 10) = 0.$$

Therefore, $x^* = 2$ and $x^* = -10$ are critical points. To determine if $x^* = 2$ is a local optimum, use $\Delta x = 0.01$ and evaluate the derivative function at $x - \Delta x$ and $x + \Delta x$.

$$\frac{dy}{dx}(x = 1.99) = (1.99)^2 + 8(1.99) - 20$$
$$= 3.9601 + 15.9200 - 20 = -0.1199.$$
$$\frac{dy}{dx}(x = 2.01) = (2.01)^2 + 8(2.01) - 20$$
$$= 4.0401 + 16.08 - 20 = 0.1201.$$

We conclude that for $x^* = 2$ the function is a local minimum since the slope of the function is less than 0 at $x - \Delta x$, equals 0 at the critical point, and is greater than 0 at $x + \Delta x$. By a similar process, we find that the function is a local maximum for $x^* = -10$. The value of the function for $x^* = -10$ and $x^* = 2$ is

$$y(-10) = 276.7 \quad \text{and} \quad y(2) = -11.3.$$

Example. Determine the values of x such that the slope of y is 0.

$$y = x^3 - 6x^2 + 9x + 10,$$
$$\frac{dy}{dx} = 3x^2 - 12x + 9 = 0.$$

Dividing through by 3 gives

$$x^2 - 4x + 3 = 0,$$

which can be factored to yield

$$(x - 3)(x - 1) = 0.$$

Thus

$$x^* = 3, \qquad x^* = 1.$$

Example. Determine the values of x such that the slope of y is 0.

$$y = -2x^3 + 4x^2 + 16x + 20,$$
$$\frac{dy}{dx} = -6x^2 + 8x + 16 = 0$$
$$-3x^2 + 4x + 8 = 0.$$

This equation cannot be factored. It is possible, however, to determine the roots of the equation through use of the quadratic formula. The quadratic formula applies to quadratic equations. The form of the quadratic equation is $cx^2 + bx + a = 0$, where c and b are the coefficients of the x^2 and x terms, and a is the constant. To determine the values of x for which the equation equals 0 (i.e., the roots of the equation), the following formula is used:

$$x = \frac{-(b) + \sqrt{b^2 - 4ac}}{2c} \quad \text{and} \quad x = \frac{-(b) - \sqrt{b^2 - 4ac}}{2c}. \tag{2.1}$$

This is more commonly written as

$$x = \frac{-(b) \pm \sqrt{b^2 - 4ac}}{2c}.$$

Comparison of the example problem and the general form of the quadratic equation shows that $c = -3$, $b = 4$, and $a = 8$. Therefore,

$$x = \frac{-4 \pm \sqrt{(4)^2 - 4(-3)(8)}}{2(-3)} = \frac{-4 \pm \sqrt{16 + 96}}{-6}$$

$$x = \frac{-4 \pm \sqrt{112}}{-6} = \frac{-4 \pm 10.58}{-6},$$

$$x = \frac{6.58}{+6} = -1.10$$

is one root, and

$$x = \frac{-14.58}{-6} = 2.43$$

is the other root. The slope of the function is 0 at $x = -1.10$ and $x = 2.43$.

Example. Determine the critical points of the following function and specify if the critical points are local optima.

$$y = 4x^3 - 15x^2 + 10x + 100,$$

$$\frac{dy}{dx} = 12x^2 - 30x + 10 = 0$$

$$\frac{dy}{dx} = 6x^2 - 15x + 5 = 0,$$

$$x = \frac{-(-15) \pm \sqrt{(15)^2 - 4(6)(5)}}{2(6)} = \frac{15 \pm \sqrt{225 - 120}}{12}$$

$$x = \frac{15 \pm \sqrt{105}}{12} = \frac{15 \pm 10.25}{12},$$

$$x = \frac{25.25}{12} = 2.10 \quad \text{and} \quad x = \frac{4.75}{12} = 0.40.$$

The critical points of the function are $x^* = 2.10$ and $x^* = 0.40$. To determine if the critical points are local optima, the derivative is evaluated at $x - \Delta x$ and $x + \Delta x$. If we specify Δx as 0.10, the calculations for determining the maximum and minimum are as follows.

For $x = 2.10$:

$$\frac{dy}{dx}(x = 2.00) = 12(2.00)^2 - 30(2.00) + 10 = -2.00$$

$$\frac{dy}{dx}(x = 2.20) = 12(2.20)^2 - 30(2.20) + 10 = +2.08.$$

For $x = 0.40$:

$$\frac{dy}{dx}(x = 0.30) = 12(0.30)^2 - 30(0.30) + 10 = +2.08$$

$$\frac{dy}{dx}(x = 0.50) = 12(0.50)^2 - 30(0.50) + 10 = -2.00.$$

The function reaches a local maximum at $x = 0.40$ and a local minimum at $x = 2.10$. The local optima are

$$Y(x = 0.40) = 4(0.40)^3 - 15(0.40)^2 + 10(0.40) + 100 = 101.9$$
$$Y(x = 2.10) = 4(2.10)^3 - 15(2.10)^2 + 10(2.10) + 100 = 91.9.$$

We can determine if a function is a local maximum or minimum from the definitions of local maximum and local minimum. A function is said to reach a local maximum at the value $x = a$ if the function evaluated at $x = a$ is greater than the function evaluated at any adjacent point. The function is a local maximum at $x^* = a$ if $f(a - \Delta x) < f(a)$ and $f(a + \Delta x) < f(a)$. The function is a local minimum if $f(a - \Delta x) > f(a)$ and $f(a + \Delta x) > f(a)$. This procedure for specifying whether a local optimum is a maximum or minimum is illustrated in the following examples.

Example. The profits of the Woodly Manufacturing Company vary with the quantity of product produced and sold according to the profit function

$$f(x) = -x^2 + 600x - 10{,}000,$$

where x represents quantity and $f(x)$ represents profits. Determine the quantity that leads to maximum profits.

The critical points of the profit function are determined by equating the derivative of the profit function with 0. Thus

$$f'(x) = -2x + 600 = 0$$

and

$$x^* = 300.$$

Profits are a maximum for Woodly Manufacturing Company when $x^* = 300$ units are produced. The profits for this output are $f(300) = \$80{,}000$. The fact that this quantity is a maximum rather than a minimum or point of inflection is verified by determining profit for $x = 299$ units and $x = 301$ units.

Example. The average costs per unit of manufacturing for Brown, Inc. varies with the percentage of capacity utilized. At low levels of output, average costs tend to be relatively high because of inefficient combination of the factors of production. Similarly, high levels of output lead to costly overtime and other inefficient combinations of the factors of production. The average cost curve for Brown, Inc. is described by the function

$$f(x) = 3500 - 16x + 0.02x^2,$$

where x represents quantity and $f(x)$ represents costs. Determine the quantity that leads to minimum average cost.

The quantity for which costs are minimum is determined by equating the derivative of the average cost function with 0.

$$f'(x) = -16 + 0.04x = 0$$

and

$$x^* = 400.$$

Costs are minimized when 400 units are produced per period. The average cost of the 400 units is $f(400) = \$300$. This is a local minimum, since $f(399)$ and $f(401)$ give higher average costs.

2.2.1 HIGHER-ORDER DERIVATIVES

We have shown that the derivative of a function, when evaluated at the point $x = a$, gives the slope of the function at the point $x = a$. It is also possible to differentiate the derivative. The derivative of the derivative is termed the second derivative. The second derivative, when evaluated at the point $x = a$, gives the slope of the first derivative at that point. This slope, as we shall see later, provides useful information in the evaluation of the original function.

A higher-order derivative is simply the derivative of a derivative. Thus, if

$$y = x^4 + 3x^3$$

is the function, then

$$\frac{dy}{dx} = 4x^3 + 9x^2$$

is the first derivative,

$$\frac{d^2y}{dx^2} = 12x^2 + 18x$$

is the second derivative,

$$\frac{d^3y}{dx^3} = 24x + 18$$

is the third derivative,

$$\frac{d^4y}{dx^4} = 24$$

is the fourth derivative, and

$$\frac{d^5y}{dx^5} = 0$$

is the fifth derivative. Each of these derivatives, when evaluated at the point $x = a$, gives the slope of the preceding derivative at that point. The first derivative describes the slope of the function, the second derivative describes the slope of the first derivative, the third derivative the slope of the second derivative, etc. In business and economic applications of differential calculus, we seldom use derivatives other than the first and second derivatives of a function. Consequently, our discussion of higher-order derivatives is limited to the first and second derivatives.

2.2.2 THE SECOND DERIVATIVE

The rules for determining second derivatives are the same as those used to determine the first derivative. The second derivative is merely the derivative of the first derivative; thus, if the first derivative is an algebraic function, then the appropriate derivative rules are applied to determine the derivative of the first derivative.

Example. $y = 3x^4 - 6x^3 + 5x^2 - 10x + 20,$

$$\frac{dy}{dx} = 12x^3 - 18x^2 + 10x - 10,$$

and

$$\frac{d^2y}{dx^2} = 36x^2 - 36x + 10.$$

Example.

$$y = \frac{(x+4)^2}{(x^2+2x)},$$

$$\frac{dy}{dx} = \frac{(x^2+2x)[2(x+4)^1] - (x+4)^2(2x+2)}{(x^2+2x)^2}$$

$$\frac{dy}{dx} = \frac{-6x^2 - 32x - 32}{(x^2+2x)^2},$$

$$\frac{d^2y}{dx^2} = \frac{(x^2+2x)^2(-12x-32) - 2(-6x^2-32x-32)(x^2+2x)(2x+2)}{(x^2+4x)^4}$$

$$\frac{d^2y}{dx^2} = \frac{-4(x^2+2x)^2(3x+8) + 8x(3x^2+16x+32)(x+2)(x+1)}{(x^2+4x)^4}.$$

2.2.3 USE OF THE SECOND DERIVATIVE IN LOCATING LOCAL OPTIMUM

The second derivative is used to identify local maxima and minima. The use of the second derivative in identification of a local minimum is illustrated in Fig. 2.6. Figure 2.6(a) shows that the function is a minimum at $x = 3$. The slope of the function is 0 at $x = 3$. In addition, the second derivative is positive, as shown in Fig. 2.6(c).

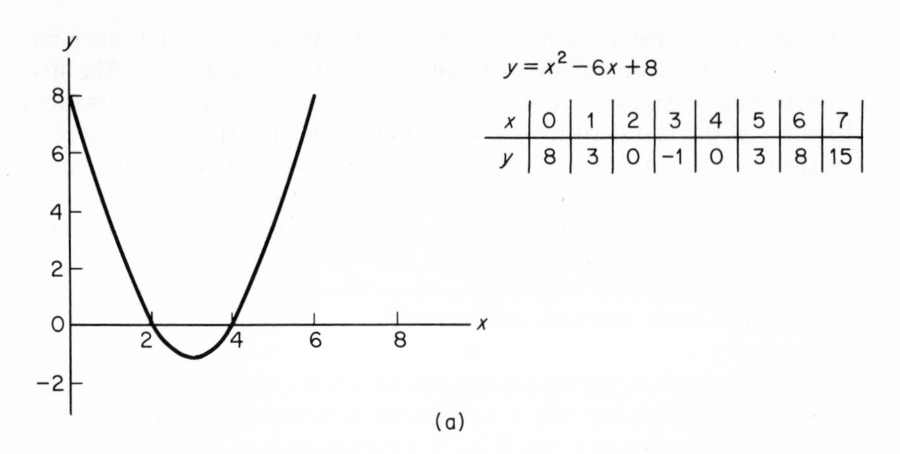

$y = x^2 - 6x + 8$

x	0	1	2	3	4	5	6	7
y	8	3	0	−1	0	3	8	15

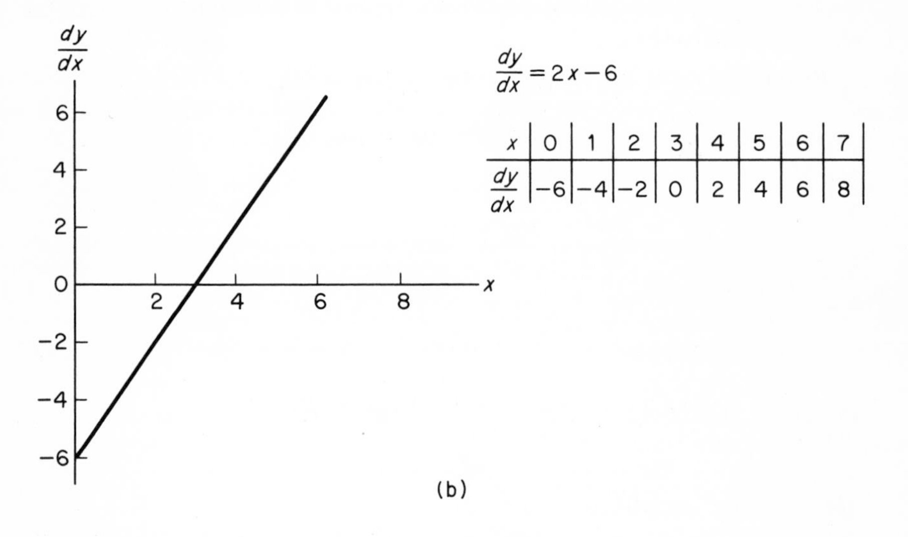

$\frac{dy}{dx} = 2x - 6$

x	0	1	2	3	4	5	6	7
$\frac{dy}{dx}$	−6	−4	−2	0	2	4	6	8

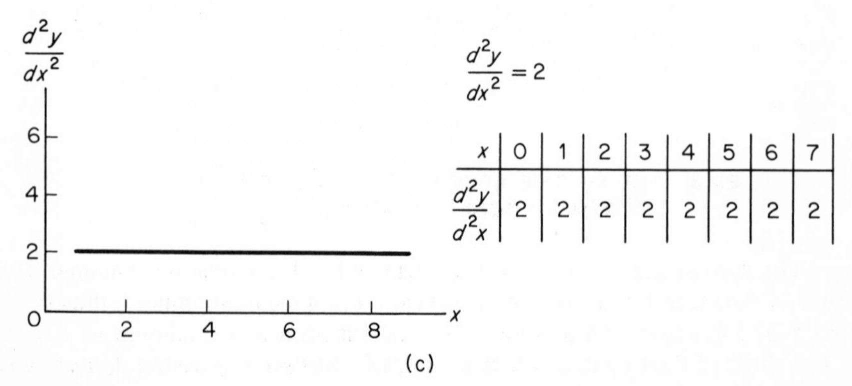

$\frac{d^2y}{dx^2} = 2$

x	0	1	2	3	4	5	6	7
$\frac{d^2y}{d^2x}$	2	2	2	2	2	2	2	2

Figure 2.6

It is not merely coincidence that the second derivative is positive when evaluated at that value of x for which the function is a local minimum. It is always true that a function is a local minimum at $x = a$ if the first derivative, when evaluated for $x = a$, is 0, and the second derivative when evaluated at $x = a$ is positive.

This principle is also illustrated in Fig. 2.7. The function is a local minimum at $x = 6$. Since $x = 6$ is a critical point, the first derivative is 0 at $x = 6$. Also, the second derivative is positive at $x = 6$. Consequently, the point $x = 6$ satisfies the dual requirement that the first derivative be 0 and the second derivative be positive.

Figure 2.7 also illustrates a local maximum. The function is a local maximum at $x = 2$. The first derivative is 0 for $x = 2$. As shown in Fig. 2.7(c), the second derivative is negative at $x = 2$. This illustrates the use of the second derivative in locating local maxima. A function reaches a local maximum at $x = a$ if the first derivative is 0 at $x = a$ and the second derivative is negative at $x = a$.

In summary, the function $f(x)$ reaches a local:

1. Maximum at $x = a$ if $f'(a) = 0$, and $f''(a) < 0$.
2. Minimum at $x = a$ if $f'(a) = 0$, and $f''(a) > 0$.

The function reaches an inflection point rather than a local maximum or minimum at $x = a$ if both first and second derivatives are 0, i.e., $f'(a) = 0$ and $f''(a) = 0$.

Example. For $f(x) = -5 + 10x - x^2$, determine the critical point. Specify whether the critical point is a maximum, a minimum or a point of inflection. Determine the value of the function for the critical point.

$$f'(x) = 10 - 2x = 0.$$

Thus $x^* = 5$ is a critical point.

$$f''(x) = -2.$$

Thus $x^* = 5$ is a local maximum.

$$f(5) = 20$$

is the value of the local maximum.

Example. For $f(x) = x^3 - 7x^2 - 5x + 20$, determine critical points and specify whether the critical points are maxima, minima, or points of inflection. Determine the value of the function for the critical points.

$$f'(x) = 3x^2 - 14x - 5 = 0$$
$$(3x + 1)(x - 5) = 0$$
$$x^* = -\tfrac{1}{3} \quad \text{and} \quad x^* = 5$$

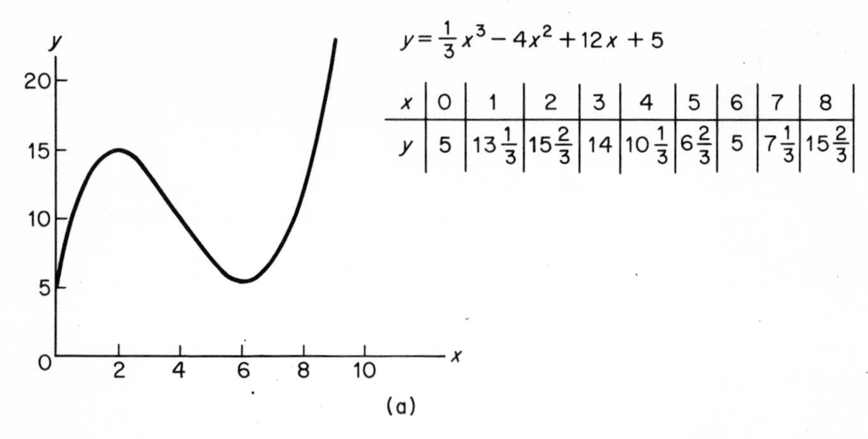

$y = \frac{1}{3}x^3 - 4x^2 + 12x + 5$

x	0	1	2	3	4	5	6	7	8
y	5	$13\frac{1}{3}$	$15\frac{2}{3}$	14	$10\frac{1}{3}$	$6\frac{2}{3}$	5	$7\frac{1}{3}$	$15\frac{2}{3}$

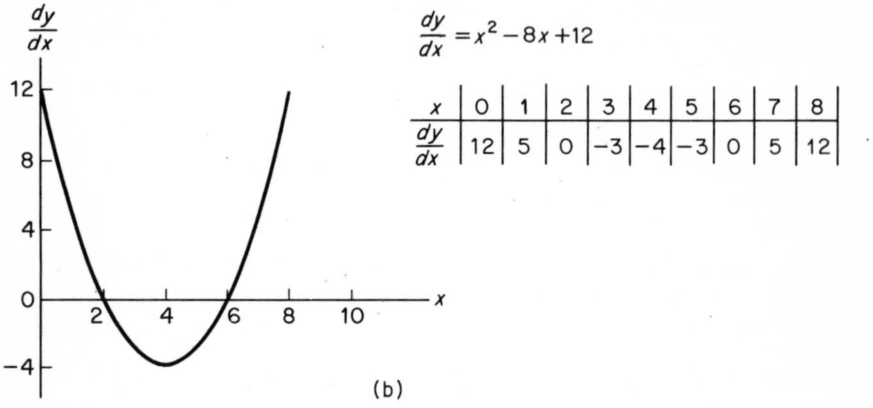

$\frac{dy}{dx} = x^2 - 8x + 12$

x	0	1	2	3	4	5	6	7	8
$\frac{dy}{dx}$	12	5	0	−3	−4	−3	0	5	12

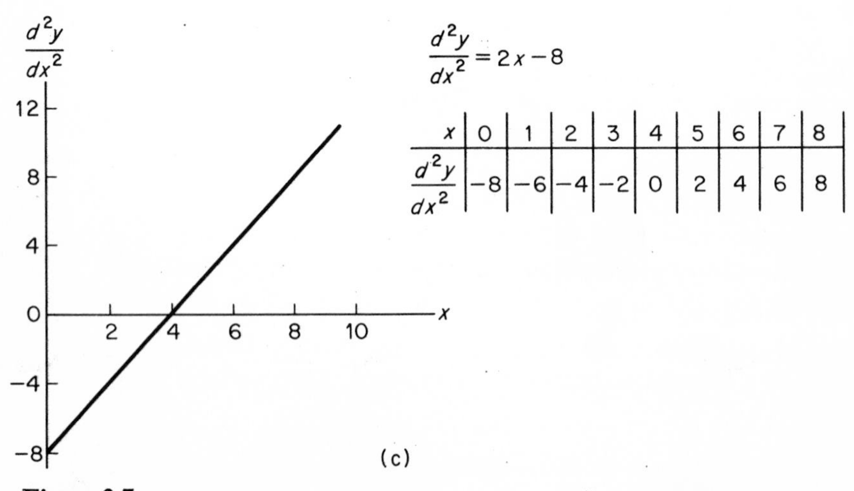

$\frac{d^2y}{dx^2} = 2x - 8$

x	0	1	2	3	4	5	6	7	8
$\frac{d^2y}{dx^2}$	−8	−6	−4	−2	0	2	4	6	8

Figure 2.7

are critical points.

$$f''(x) = 6x - 14$$
$$f''(-\tfrac{1}{3}) = -16.$$

Thus $x = -\frac{1}{3}$ is a maximum.

$$f''(5) = +16.$$

Thus $x = 5$ is a minimum.

$$f(-\tfrac{1}{3}) = 20.85$$

is the value of the local maximum.

$$f(5) = -55$$

is the value of the local minimum.

Example. For $f(x) = -6x^3 + 2x^2 + 14x - 20$, determine the critical points and specify whether the critical points are maxima, minima, or points of inflection. Determine the value of the function at the critical points.

$$f'(x) = -18x^2 + 4x + 14 = 0$$
$$x = \frac{-4 \pm \sqrt{16 - 4(-18)(14)}}{2(-18)} = \frac{-4 \pm \sqrt{1024}}{-36} = \frac{-4 \pm 32}{-36}$$

and $x = 1$ and $x = -\frac{7}{9}$ are the critical points.

$$f''(x) = -36x + 4$$
$$f''(1) = -32.$$

$x^* = 1$ is a maximum.

$$f''(-\tfrac{7}{9}) = +32.$$

$x^* = -\frac{7}{9}$ is a minimum

$$f(1) = -10$$

is the value of the local maximum.

$$f(-\tfrac{7}{9}) = -26.9$$

is the value of the local minimum.

Example. For $f(x) = 0.01x^3 - 0.15x^2 + 0.50x + 5$, determine critical points and specify whether the critical points are maxima, minima, or points of inflection. Determine the value of the function at the critical points.

$$f'(x) = 0.03x^2 - 0.30x + 0.50 = 0$$
$$x = \frac{-(-0.30) \pm \sqrt{(-0.30)^2 - 4(0.03)(0.50)}}{2(0.03)}$$
$$x = \frac{0.30 \pm \sqrt{0.09 - 0.06}}{0.06} = \frac{0.30 \pm \sqrt{0.03}}{0.06} = \frac{0.30 \pm 0.1732}{0.06}$$

Thus

$$x^* = \frac{0.4732}{0.06} = 7.9,$$

and

$$x^* = \frac{0.1268}{0.06} = 2.1$$

are the solutions (roots) of the quadratic equation.

$$f''(x) = 0.06x - 0.30,$$

and

$$f''(2.1) = -0.174.$$

$x^* = 2.1$ is a maximum.

$$f''(7.9) = +0.174,$$

$x^* = 7.9$ is a minimum.

The value of the function at 2.1 and 7.9 is

$$f(2.1) = 5.48 \quad \text{and} \quad f(7.9) = 4.52.$$

2.2.4† MEANING OF THE SECOND DERIVATIVE

The second derivative gives the rate of change of the slope of the original function. The reader should note that *the rate of change of the slope of the original function* differs from the *rate of change* (or slope) *of the original function.* The rate of change of the original function is given by the first derivative, whereas the rate of change of the slope of the original function is given by the second derivative.

The relationship between the function, the first derivative, and the second derivative is described below and illustrated in Table 2.1. A knowledge of these relationships is useful in understanding the behavior of functions.

Referring to Table 2.1, a function that is increasing at an increasing rate is a function whose slope is positive and is becoming more positive for larger values of the independent variable. A function that is increasing at a decreasing rate is a function whose slope is positive but is becoming less positive for larger values of x. Similarly, a function that is decreasing at an increasing rate is a function whose slope is negative and becoming more negative for larger values of x. A function that is decreasing at a decreasing rate is a function whose slope is negative and becoming less negative for larger values of x.

---

†This section can be omitted without loss of continuity with later sections.

A point of inflection of a function occurs when the second derivative equals zero. This occurs at that value of the independent variable for which the function changes from (1) increasing at a increasing rate to increasing at a decreasing rate, (2) increasing at a decreasing rate to increasing at an increasing rate, (3) decreasing at a decreasing rate to decreasing at an increasing rate, and (4) decreasing at an increasing rate to decreasing at a decreasing rate. Illustrations of these four points of inflection are shown in Table 2.1.

As an example of the relationships described in the table, consider again the function graphed in Fig. 2.6(a). The quadratic function decreases as x approaches 3, reaches a minimum at $x^* = 3$, and increases for subsequent increases in x. The slope of the function is plotted in Fig. 2.6(b). This graph shows that the function has a negative slope for values of x less than 3, a slope of 0 at the point $x = 3$, and a positive slope for values of x greater than 3.

The second derivative is graphed in Fig. 2.6(c). This graph shows that the rate of change of the slope of the original function is $\frac{d^2y}{dx^2} = 2$. We interpret this to mean that the slope of the original function is becoming less negative as x approaches the critical value $x^* = 3$ and more positive for values of x greater than $x^* = 3$. Alternatively, we can state that a positive second derivative indicates that the function is decreasing but at a decreasing rate for values of x less than 3 and increasing at an increasing rate for values of x greater than 3. The function is a minimum when the first derivative is 0 and the second derivative is positive.

The relationship between the function, the first derivative, and the second derivative is illustrated for a cubic function in Fig. 2.7. The function, graphed in Fig. 2.7(a), is a local maximum at $x^* = 2$ and a local minimum at $x^* = 6$. The slope of the function at both critical points is 0. This is illustrated by the graph of the slope of the function in Fig. 2.7(b).

The second derivative is graphed in Fig. 2.7(c). The second derivative is negative for values of x less than 4 and positive for values of x greater than 4. The function increases, but at a decreasing rate of increase for $0 \leq x < 2$. The function reaches a maximum at $x^* = 2$, and decreases at an increasing rate of decrease for $2 < x \leq 4$. The function reaches a point of inflection at $x = 4$ and for $4 < x < 6$ the function continues to decrease but at a decreasing rate of decrease. The minimum point is reached at $x^* = 6$. For $x > 6$ the function increases at an increasing rate of increase. This description of the behavior of the function is possible either by inspection of the function or by inspection of the first and second derivatives.

Example. Sketch the function whose first and second derivatives are

$$f'(x) = 2x^2 - 12x + 16,$$

TABLE 2.1

Value of First Derivative	*Value of Second Derivative*	*Description of the Function*	*General Appearance*
positive	positive	increasing at an increasing rate	 $f'(a) = +$ $f''(a) = +$
positive	negative	increasing at a decreasing rate	 $f'(a) = +$ $f''(a) = -$
zero	positive	local minimum	 $f'(a) = 0$ $f''(a) = +$
zero	negative	local maximum	 $f'(a) = 0$ $f''(a) = -$
negative	positive	decreasing at a decreasing rate	 $f'(a) = -$ $f''(a) = +$
negative	negative	decreasing at an increasing rate	 $f'(a) = -$ $f''(a) = -$
Points of Inflection			
1. positive	zero	point of inflection	 $f'(a) = +$ $f''(a) = 0$
2. positive	zero	point of inflection	 $f'(a) = +$ $f''(a) = 0$
3. negative	zero	point of inflection	 $f'(a) = -$ $f''(a) = 0$
4. negative	zero	point of inflection	 $f'(a) = -$ $f''(a) = 0$
5.† zero	zero	point of inflection	$f'(a) = 0$ $f''(a) = 0$ or $f'(a) = 0$ $f''(a) = 0$

†Functions of the type $y = x^4$ are special cases since these functions reach a minimum at $x = 0$ (first derivative is zero at $x = 0$) and their second derivative is zero at this minimum, yet the function reaches a minimum rather than a point of inflection at $x = 0$.

and

$$f''(x) = 4x - 12.$$

The critical points are obtained by setting the first derivative equal to 0 and solving for the roots of the equation.

$$2x^2 - 12x + 16 = 0$$
$$2(x - 4)(x - 2) = 0.$$

Thus $x^* = 2$ and $x^* = 4$.

Equating the second derivative with 0 shows the inflection point to be $x = 3$. At $x^* = 2$ the second derivative is negative. Therefore, $x^* = 2$ is a local maximum. Similarly, $x^* = 4$ is a local minimum. Based upon this information, the function can be sketched as follows.

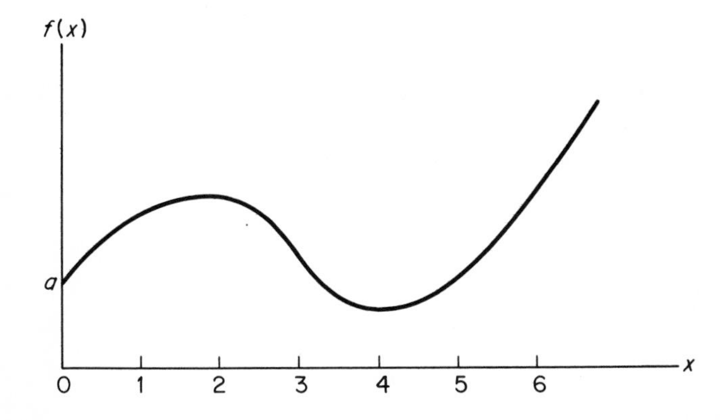

As shown in the preceding example, it is possible to sketch a function if we are given the first derivative of the function. It is not possible, however, to determine the original position of the function on the graph. Thus, in the preceding example we show the value of the function when $x = 0$ as a. The value of the intercept a is lost in taking the derivative of the function, since the derivative of any constant is 0. The general shape of the function can be sketched from the derivatives, but we must have additional information to determine the original position of the function on the vertical axis.

Example. Verify that the two functions $f_1(x) = x^2 - 6x + 6$ and $f_2(x) = x^2 - 6x + 4$ have the same general shape.

The two functions are identical with the exception that the value of the dependent variable in the first function is two units greater than the value of the dependent variable in the second function. Since both functions have the same shape, the slope of the functions at any value of the independent variable is the same. This is verified by the fact that the derivatives of the functions are identical, i.e., $f'(x) = 2x - 6$.

2.3 absolute and local maximum and minimum

In the first section of this chapter we discussed both absolute and local maxima and minima. Local maxima and minima can be determined from the first and second derivatives. The absolute maxima and minima can be found only by comparing the local maxima and local minima with the value of the function at the end points, and selecting the absolute maximum and minimum. This procedure is illustrated by the following two examples.

Example. Determine absolute and local maximum and minimum for

$$f(x) = \frac{x^3}{3} - 4x^2 + 12x + 20, \qquad \text{for } 0 \leq x \leq 10.$$

Local maximum and minimum are determined from the first and second derivatives.

$$f'(x) = x^2 - 8x + 12 = 0$$
$$f'(x) = (x - 2)(x - 6) = 0.$$
$$x^* = 2 \quad \text{and} \quad x^* = 6$$

are critical points.

$$f''(x) = 2x - 8$$
$$f''(2) = -4.$$

Therefore, $x^* = 2$ is a local maximum.

$$f''(6) = +4.$$

Therefore, $x^* = 6$ is a local minimum.

The absolute maximum and minimum are found by evaluating the function at the critical points and the end points.

$$f(0) = \frac{(0)^3}{3} - 4(0)^2 + 12(0) + 20 = 20.0$$

$$f(2) = \frac{(2)^3}{3} - 4(2)^2 + 12(2) + 20 = 30.7$$

$$f(6) = \frac{(6)^3}{3} - 4(6)^2 + 12(6) + 20 = 20.0$$

$$f(10) = \frac{(10)^3}{3} - 4(10)^2 + 12(10) + 20 = 73.3.$$

The absolute minimum is $f(x) = 20.0$, which occurs when $x = 2$ and $x = 6$, and the absolute maximum is $f(x) = 73.3$, which occurs at $x = 10$.

Example. Determine absolute and local minimum and maximum for

$$f(x) = 10 + 6x - x^2 \qquad \text{for } 0 \leq x \leq 5.$$
$$f'(x) = 6 - 2x = 0.$$

Therefore $x^* = 3$ is a critical point.

$$f''(3) = -2,$$

and the function is a local maximum, since $f''(3) < 0$.

$$f(0) = 10$$
$$f(3) = 19$$
$$f(5) = 15.$$

Thus, $x = 0$ is the absolute minimum and $x = 3$ is the absolute maximum.

Example. Determine the optima for $f(x) = x^3 - 3x^2 - 72x + 16$.

$$f'(x) = 3x^2 - 6x - 72 = 0$$
$$f'(x) = 3(x^2 - 2x - 24) = 0$$
$$f'(x) = 3(x - 6)(x + 4) = 0.$$
$$x^* = 6 \quad \text{and} \quad x^* = -4$$

are critical points.

$$f''(x) = 6x - 6.$$
$$f''(-4) = -30,$$

and the critical point $x = -4$ is a local maximum.

$$f''(6) = 30,$$

and the critical point $x = 6$ is a local minimum.

Since the domain is not specified in this example, it is assumed that $-\infty < x < \infty$. The convention, when the domain is not specified, is to specify only local optima. In this example, the function reaches a local maximum of $f(-4) = 192$ at $x = -4$ and a local minimum of $f(6) = -308$ at $x = 6$.

In summary, a function can reach a local optimum at an interior value of the independent variable, that is, a value of the independent variable that is not an end point of the domain. A function can reach an absolute maximum or minimum at either an interior point or an end point. If, however, the end points are not defined, absolute maximum or minimum are not specified.

problems

1. Determine critical points for the following functions and specify whether the function is a maxima, minima, or points of inflection for the critical points.

   a. $f(x) = x^2 - 6x + 3$
   b. $f(x) = 1.5x^2 + 9x + 12$
   c. $f(x) = 2x^2 - 12x + 10$
   d. $f(x) = -3x^2 + 7x - 12$
   e. $f(x) = \frac{x^3}{3} - \frac{5x^2}{2} + 6x + 12$
   f. $f(x) = \frac{2x^3}{3} - \frac{x^2}{2} - 6x$
   g. $f(x) = x^3 - 3x + 6$
   h. $f(x) = \frac{x^3}{3} + 7x^2 + 40x + 100$
   i. $f(x) = \frac{x^3}{3} - \frac{3x^2}{2} - 6x + 12$
   j. $f(x) = \frac{2x^3}{3} - 3x^2 + 3x + 6$
   k. $f(x) = x^3 + \frac{7x^2}{2} - 6x + 10$
   l. $f(x) = \frac{2x^3}{3} - x^2 - 24x + 24$

2. Find all local and absolute maxima and minima for the following functions.

   a. $f(x) = x^2 - 8x + 6$ for $0 \leq x \leq 6$
   b. $f(x) = -x^2 + 12x - 10$ for $0 \leq x \leq 10$
   c. $f(x) = \frac{-x^3}{3} + 5x^2 - 6x + 10$ for $0 \leq x \leq 20$
   d. $f(x) = 2x + 3$ for $0 \leq x \leq 5$
   e. $f(x) = \frac{-2x^3}{3} - 4x^2 + 6x + 10$ for $-10 \leq x \leq 5$
   f. $f(x) = \frac{-x^3}{3} - 4x^2 - 4x + 6$ for $-20 \leq x \leq 0$

3. Sketch the function whose first derivative is the following.

   a. $f'(x) = x - 3$
   b. $f'(x) = -2x + 2$
   c. $f'(x) = x^2 - 8x + 4$
   d. $f'(x) = 2x^2 - 20x + 6$
   e. $f'(x) = -x^2 - 12x - 3$
   f. $f'(x) = -3x^2 - 15x + 4$
   g. $f'(x) = 3$
   h. $f'(x) = x^2 - 16x$

4. The profit function of the Clark Manufacturing Company is described by a quadratic function. The following data points were obtained from accounting records at Clark: output of 12,000 units resulted in profit of \$120,000; output of 15,000 units resulted in profit of \$130,000; output of 18,000 units resulted in profit of \$128,000. Determine the quadratic function that describes profit as a function of the number of units produced and determine that level of output which results in maximum profit.

5. The cost of manufacturing a certain assembly at Clark Manufacturing has been found to be described by the quadratic function. Three data points from the cost curve are the following: The average cost per unit is 80¢ when output is 600 units; the average cost is 65¢ when output is 700 units; the average cost is 70¢ when output is 800 units. Determine the quadratic function that describes the functional relationship between average cost and output, and determine the output that leads to minimum average cost.
6. It is known that the earning ability of an individual increases as he grows older and then decreases as he approaches retirement age. In a certain profession average earnings at age 40 are $25,000. This figure increases to $32,000 by age 55 and declines to $28,000 at age 60. Determine the quadratic function that describes the relationship between age and earnings, and determine the age that leads to maximum earnings.
7. The average cost per unit of output at Cleveland Electronics is $140 for the 200th unit, $125 for the 250th unit, and $130 for the 275th unit. Determine the quadratic function that describes the relationship between average cost per unit and output, and find that unit of output which has minimum average cost.
8. Sales of a certain product are known to be related to advertising expenditures. In a controlled experiment in which factors such as price, quality, etc., were held constant and advertising expenditures were varied, the following results were observed: Advertising expenditures of $100 resulted in sales of $10,000; advertising expenditures of $300 resulted in sales of $11,500; advertising expenditures of $500 resulted in sales of $12,500. Determine the function that describes the relationship between advertising and sales, and determine the advertising expenditures that result in maximum sales.
9. The profit of Merril Equipment Company is determined by the number of units of equipment rented by Merril. Profits of $1000 are possible when 75 units are rented; profits of $1200 are possible when 85 units are rented; profits of $1300 are possible with the rental of 100 units of equipment. Determine the functional relationship between profit and the number of units of equipment rented, and determine the number of rental units that make possible maximum profits.

3

single variable business and economic models

Chapters 1 and 2 were concerned primarily with the introduction of the concepts and methodology of differential calculus. Chapter 3 provides some of the more important business and economic applications of differential calculus. In Chapter 3 pricing and output decision of a firm are examined through the use of differential calculus. Two production models are also considered in this chapter, the basic economic order quantity model and the production lot size model.

3.1 profit maximization

One of the important applications of differential calculus is in the development of models that describe pricing and output decisions in the business firm. The basic assumption underlying the majority of these models is maximization of profit. The firm is assumed to produce the quantity of output for which profits are maximized. Profits of a firm depend upon both revenue and cost. The revenue of a firm is a function of the price of its products and the quantity sold. The cost of production to the firm depends upon the costs of inputs used in the production process and the quantity of product manufactured. Profits are, of course, the difference between revenue and costs.

We shall consider revenue and cost functions separately in the following two sections. They are then combined into a model of profit maximization.

3.1.1 REVENUE

The revenue generated by a firm from the sale of a product depends upon the price of the product and the quantity of the product sold. The price of the product and the quantity of the product sold are related variables. The relationship between these variables is described by the demand function.

DEMAND FOR A PRODUCT. The demand function for a product is the functional relationship between the quantity of the product offered for sale and the price that this quantity will bring in the market. The *law of demand* describes this functional relationship: The quantity of a product purchased during a specified period of time will be larger the lower the price, *ceteris paribus*.† The ceteris paribus factors include tastes and incomes of the individuals purchasing the product and the prices of products that are substitutes and complements. If these ceteris paribus conditions remain constant, then the demand function for a product plots as shown in Fig. 3.1. In this figure we assume that the price is a function of the quantity offered for sale; therefore, price is plotted on the vertical axis and quantity on the horizontal axis. The rationale behind this assumption is that a certain quantity of goods is brought to the market place and the price is established based upon this quantity. The larger the quantity offered for sale, the smaller the price obtainable for the goods.

It is equally justifiable to treat quantity as the independent variable and price as the dependent variable. With a moment's reflection, one realizes that a change in price should result in a change in quantity; or, alternatively, a change in quantity should lead to a change in price. Thus, each variable is dependent upon the other. It is, therefore, the analyst's choice as to which variable is treated as dependent. Most students who have taken an economics course are familiar with diagrams, such as Fig. 3.1, which show price as the dependent variable and quantity the independent variable. We shall follow

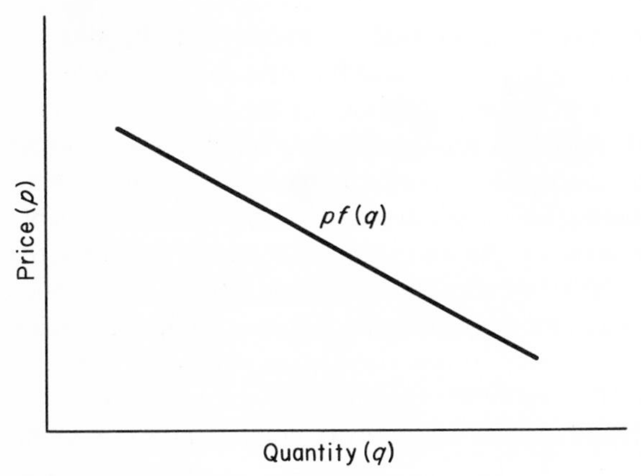

Figure 3.1

†*Ceteris paribus* means that all other relevant factors remain unchanged.

the custom established by the economist and treat price as the dependent variable in the demand function.

Changes in one or more of the ceteris paribus conditions are shown by shifts in the demand function. As an example, if incomes of individuals purchasing the product increase, the demand function might shift from position A in Fig. 3.2 to position B. Similarly, should the price of a competing product be reduced, the demand function could possibly shift to position C in Fig. 3.2.

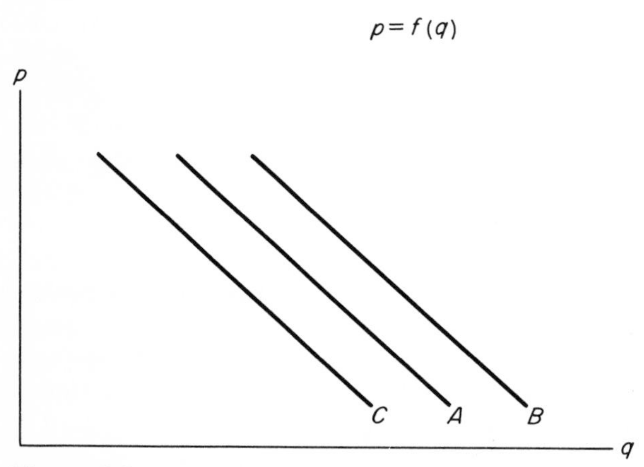

Figure 3.2

Demand for the Output of a Firm. The demand function for a product describes the relationship between the demand for various quantities of the product and price of the product. The demand function for the product is not normally equivalent to the demand function for the production of an individual firm. If the individual firm produces only a small segment of the total production of the product, then it might be possible for the firm to increase or decrease the quantity of the product produced without noticeably affecting the price of the product. As an example, if we assume that the demand function in Fig. 3.3(a) describes the relationship between the price and quantity of wheat grown in the United States, then it is not unreasonable to assume that the demand function for the wheat of an individual farmer is as shown in Fig. 3.3(b). In this figure we again assume that the quantity of wheat determines the price, and therefore quantity is plotted on the horizontal axis with price on the vertical axis. If q^* is produced and brought to market, the price will be p^*. The individual farmer can sell his entire crop for p^*, but he would not be able to raise the price by withholding a portion of his crop. This is the case of *pure competition*.

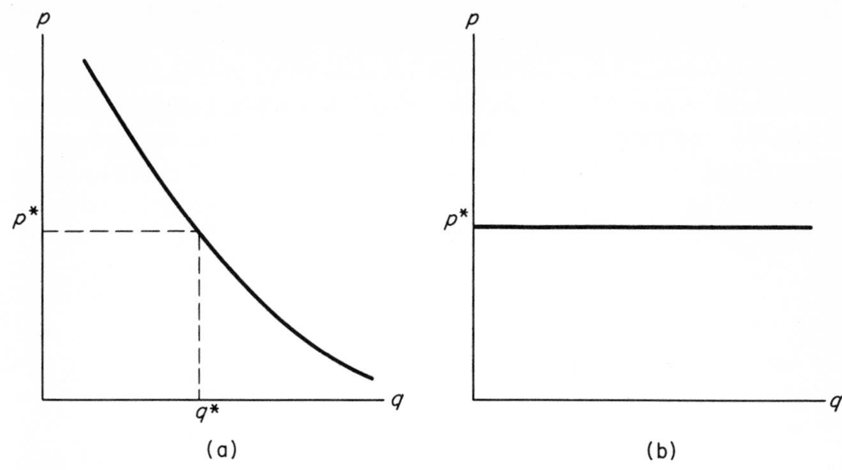

Figure 3.3

There are a number of firms that produce the entire quantity of a product offered for sale in the market. This situation of a single manufacturer occurs when the product is protected, for example, by patents or copyrights, or alternatively, when barriers restrict other firms from engaging in production of the product. The demand for the production of these firms is described by the demand function for the product (Fig. 3.1). This is the case of *pure monopoly*.

The large majority of firms compete in markets that fall in between the extremes of pure competition and pure monopoly. These firms have some control over their price, although if they reduce their price they can expect their competitors to follow with similar reductions. Quantities produced by the individual firms can vary over a limited range without price changes, although again the firm would find it difficult to increase sales significantly without lowering the price of the output. The demand for the production of these firms is described by a negatively sloping demand function similar to that shown in Fig. 3.1. This is the case of *oligopolistic* or *monopolistic competition*.

TOTAL REVENUE. A firm's total revenue that is derived from the sale of a product is given by *price multiplied by quantity*. Expressed mathematically:

$$TR = p \cdot q. \tag{3.1}$$

From the demand function we observe that price is a function of quantity:

$$p = f(q). \tag{3.2}$$

Total revenue is thus represented as a function of quantity:

$$TR = f(q) \cdot q. \tag{3.3}$$

This function is shown in Fig. 3.4 for two demand functions. It is assumed for both demand functions that the relationship between price and quantity can be approximated by a continuous function.

The total revenue function shown in Fig. 3.4(a) is based upon a negatively sloping demand function. For this case, if additional units of the product are to be offered for sale during the time period, the price of the product could be expected to fall. This results in an increase in total revenue for values of q less than q^*, and a decrease in total revenue for values of q greater than q^*. The total revenue function shown in Fig. 3.4(b) increases as q increases, since p is constant.

MARGINAL REVENUE. One of the useful measures of revenue is marginal revenue. The marginal revenue is the change in total revenue for a unit change in the quantity offered for sale. Using the symbol Δ to represent change, the marginal revenue can be represented by

$$MR = \frac{\Delta TR}{\Delta q}.$$

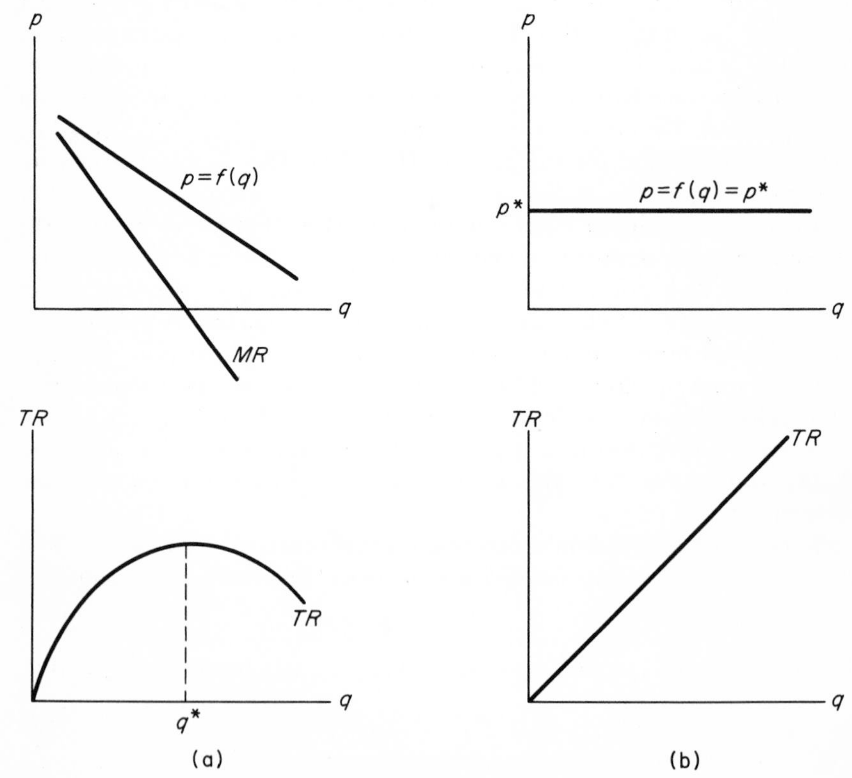

Figure 3.4

For those cases in which the demand and total revenue functions are continuous, the expression $MR = \frac{\Delta TR}{\Delta q}$ is replaced by the expression $MR = \lim_{\Delta q \to 0} \frac{\Delta TR}{\Delta q}$. The expression $\lim_{\Delta q \to 0} \frac{\Delta TR}{\Delta q}$ is the derivative of total revenue with respect to q. The derivative of the total revenue function thus describes the change in total revenue when an additional unit of product is offered for sale in the market place.

The marginal revenue is given by

$$MR = \frac{d(TR)}{dq} = f(q) + q[f'(q)]. \tag{3.4}$$

The marginal revenue function is shown to be the sum of the demand function and q times the derivative of the demand function. In Fig. 3.4(a), the demand function has a negative slope, and therefore the derivative of the demand function is negative. Thus, the marginal revenue function plots to the left of the demand function. In Fig. 3.4(b), the derivative of the demand function is 0. Thus, in the case of pure competition the marginal revenue and demand functions are the same.

Again referring to Fig. 3.4(a), note that revenue is a maximum when marginal revenue is 0. The quantity which produces maximum revenue can be determined by equating the marginal revenue function with zero and solving for the critical value q^*.

AVERAGE REVENUE. The average revenue from a product is found by dividing the total revenue by the quantity of the product sold. The function that describes average revenue is the quotient of the total revenue function and the quantity. From (3.5) we see that average revenue function and the demand function are equivalent.

$$AR = \frac{TR}{q} = \frac{f(q)q}{q} = f(q), \tag{3.5}$$

PRICE ELASTICITY OF DEMAND. The price elasticity of demand provides a measure of the effect of a change in price upon total revenue. If we represent the price elasticity by E, the formula for price elasticity is

$$E = \frac{\frac{\Delta q}{q}}{\frac{\Delta p}{p}}. \tag{3.6}$$

From Formula (3.6) it can be seen that the elasticity of demand is simply the ratio of the percentage change in quantity divided by the percentage change in price. By algebraic manipulation rewrite the formula for elasticity as

$$E = \frac{\Delta q}{\Delta p} \cdot \frac{p}{q}. \tag{3.7}$$

For continuous functions $\Delta p/\Delta q$ is equivalent to dp/dq. Thus, the elasticity for a continuous function is

$$E = \frac{dq}{dp} \cdot \frac{p}{q}. \tag{3.8}$$

The expression $\frac{dq}{dp}$ in formula (3.8) is the reciprocal of the slope of the demand function. Since the slope of the demand function is negative, the elasticity of demand is negative.

The price elasticity of demand gives the effect upon total revenue of a change in the price of a product. For $0 > E > -1$, a decrease in price results in a less than proportional increase in quantity, and thus a decrease in total revenue. Similarly, for $0 > E > -1$, an increase in price results in an increase in total revenue. For $E = -1$, a change in price results in a proportional change in quantity, and total revenue remains constant. For $E < -1$, a decrease in price results in a greater than proportional increase in quantity demanded and an increase in total revenue. Similarly, an increase in price results in a decrease in total revenue.

It is customary to consider only the absolute value of the elasticity. If we use this convention, for $0 < E < 1$ a decrease in price results in a decrease in total revenue. Similarly, an increase in price results in an increase in total revenue. This is termed *inelastic demand.* For $E = 1$, a decrease in price is counterbalanced by an increase in quantity demanded. This is termed *unit elasticity.* For $E > 1$, a decrease in price is accompanied by an increase in total revenue, and an increase in price is accompanied by a decrease in total revenue. This is termed an *elastic demand.*

Example. As an example of the concepts discussed in Sec. 3.1.1, consider the Hess Electronics Company, which manufactures an electronic module used in television sets. On the basis of market studies, they have found that the quantity of modules sold varies with the price of the module. Specifically, the demand schedule is as follows:

Price per unit	*Units sold (thousands)*
$1.00	5.0
1.25	4.5
1.50	4.0
1.75	3.5
2.00	3.0
2.25	2.5

Based upon the schedule of demand, the demand function is determined. If we assume a linear demand function, the functional relationship between price and quantity is determined by selecting two data points and establishing the linear function.†

$$p = f(q) = 3.50 - 0.50q.$$

Total revenue is the product of price and quantity:

$$TR = p \cdot q = (3.50 - 0.50q)q = 3.50q - 0.50q^2.$$

Marginal revenue is the derivative of total revenue with respect to quantity:

$$MR = \frac{d(TR)}{dq} = 3.50 - 1.00q.$$

Average revenue is found by dividing total revenue by q:

$$AR = \frac{TR}{q} = 3.50 - 0.50q.$$

The quantities TR, AR, MR, and demand are plotted in Fig. 3.5. Note that marginal revenue is 0 when total revenue is maximum.

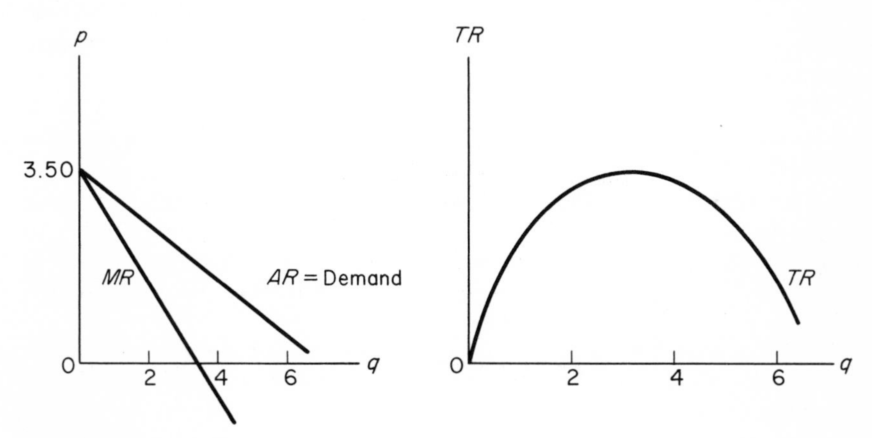

Figure 3.5

The elasticity of demand can be determined for various price levels. Since $p = f(q)$ and $q = h(p)$ are inverse functions, we can determine dq/dp by using the rule for determining derivatives of inverse functions (Rule 10). Thus

$$dq/dp = \frac{1}{dp/dq} = \frac{1}{-0.50} = -2.$$

---

†The reader may wish to review the method of establishing a linear function through two data points discussed in Appendix A.

The elasticity, by formula (3.8), is

$$E = \frac{dq}{dp} \cdot \frac{p}{q} = \frac{-2p}{q}.$$

The absolute value of the elasticity is

$$E = \frac{2p}{q}.$$

The elasticity of demand for alternative values of p and q is determined in the following manner:

Inelastic:

$$0 < \frac{2p}{q} < 1,$$

or, alternatively,

$$0 < 2p < q.$$

Unit Elastic:

$$\frac{2p}{q} = 1,$$

or, alternatively,

$$2p = q.$$

Elastic:

$$\frac{2p}{q} > 1,$$

or, alternatively,

$$2p > q.$$

The Hess Electronics Company provided an example of the important revenue curves and the elasticity of demand. Although the demand and revenue information was available in the example problem, we did not attempt to make a decision upon pricing and output. The reason, of course, was because the costs involved in the production of the output must also be considered.

In those cases in which all costs are fixed, it is possible to determine the price and quantity that lead to maximum profits. In this case, we merely maximize total revenue. As an example, consider the following problem.

Example. There are two highways between Kansas City, Missouri, and Wichita, Kansas, the Kansas Turnpike and U.S. Highway 50. The average two-way traffic between these two cities on the turnpike is 1000 cars. It has been estimated that approximately 500 cars travel between these cities using Highway 50. The toll for using the turnpike is \$4.00. If we assume that all traffic would use the turnpike if the toll were reduced to \$2.00 and furthermore that the demand function for the turnpike is linear, determine the price and quantity that lead to maximum revenue for the turnpike.

We have two points on the demand curve: when $p = \$4.00$, $q = 1000$; and when $p = \$2.00$, $q = 1500$. Following the custom of treating price as the dependent variable and quantity as the independent variable and using the method outlined in Sec. A.1 of Appendix A, we obtain the linear demand function

$$p = 8.00 - 0.004q.$$

Total revenue is the product of price and quantity.

$$TR = pq = (8.00 - 0.004q)q = 8.00q - 0.004q^2.$$

Total revenue is a maximum when marginal revenue is 0.

$$\frac{d(TR)}{dq} = MR = 8.00 - 0.008q = 0$$

$$q^* = 1000.$$

From our analysis and on the basis of the assumption of a linear demand curve, we conclude that revenue is a maximum at the current price of \$4.00 and quantity of 1000 cars.

3.1.2 COST

The total cost incurred by a firm in the production of a product is customarily assumed to consist of fixed costs and variable costs. Fixed costs are those costs that remain constant as output varies. These include items such as interest, insurance, property taxes, and salaries of those people necessary even during periods of temporary shutdown of production. Variable costs are those that vary with the volume of output. These include direct labor, cost of materials, and other expenses directly incurred by the production process.

The relationship between total cost and output is described by the cost function. Economists normally consider two cost functions—the short-run cost function and the long-run cost function. In the short run, certain inputs to the production process are fixed in amount, and a firm can vary production only by increasing or decreasing the variable inputs. In the long run all inputs are variable. Thus in the short run the production of a product is limited by fixed plant and equipment, whereas in the long run the firm is able to vary the quantity of plant and equipment. We shall consider the short-run cost curves in this text. Long-run curves are quite similar in appearance to short-run cost curves. The student can, therefore, quite easily extend our discussion to incorporate the differences between short-run and long-run curves with the aid of an economics text on price theory.

CONVENTIONAL COST FUNCTIONS. The general form of the total cost function for a firm is described by the cubic function

$$TC = f(q) = a + bq + cq^2 + dq^3, \tag{3.9}$$

TFC TVC.

where a, b, c, and d are parameters and q is the quantity of output. Since total costs are composed of total fixed costs and total variable costs, the function in (3.9) is often stated as

$$TC = TFC + TVC, \tag{3.10}$$

and it can therefore be seen that

$$TFC = a, \tag{3.11}$$

and

$$TVC = bq + cq^2 + dq^3. \tag{3.12}$$

The conventional short-run total cost curve is shown in Fig. 3.6(a). Those costs incurred when the quantity produced is 0 are total fixed costs. As output expands, the proportion of variable inputs and fixed inputs to the production process is altered. This alteration in proportions typically results in more efficient and less costly production. This is shown by the *TC* and *TVC* curves. Both curves show that total costs increase; however, the rate of increase in total cost is decreasing. The proportion of variable inputs and fixed inputs continues to change as production expands. After reaching a certain level of production, the proportion of variable and fixed inputs to the production process passes an optimum. This is reflected in the total cost curve by total cost increasing at an increasing rate of increase.

SHORT-RUN COST CURVES. The appropriate functional relationships for the average and marginal cost curves can be determined from the total cost curve. Since total cost is a cubic function, marginal cost will be a quadratic function.

$$MC = \frac{d(TC)}{dq}. \tag{3.13}$$

The average cost curves are found by dividing the total cost curves by q. Thus,

$$ATC = \frac{TC}{q} = dq^2 + cq + b + \frac{a}{q}, \tag{3.14}$$

$$AVC = \frac{TVC}{q} = dq^2 + cq + b, \tag{3.15}$$

$$AFC = \frac{TFC}{q} = \frac{a}{q}. \tag{3.16}$$

MC and *AVC* are quadratic functions. *AFC* takes the form of what mathematicians term a rectangular hyperbola. The sum of average variable and average fixed cost equals average total cost. That is,

$$ATC = AVC + AFC. \tag{3.17}$$

The relationship among marginal cost, average variable cost, average total cost, and average fixed cost is diagrammed in Fig. 3.6(b). This figure

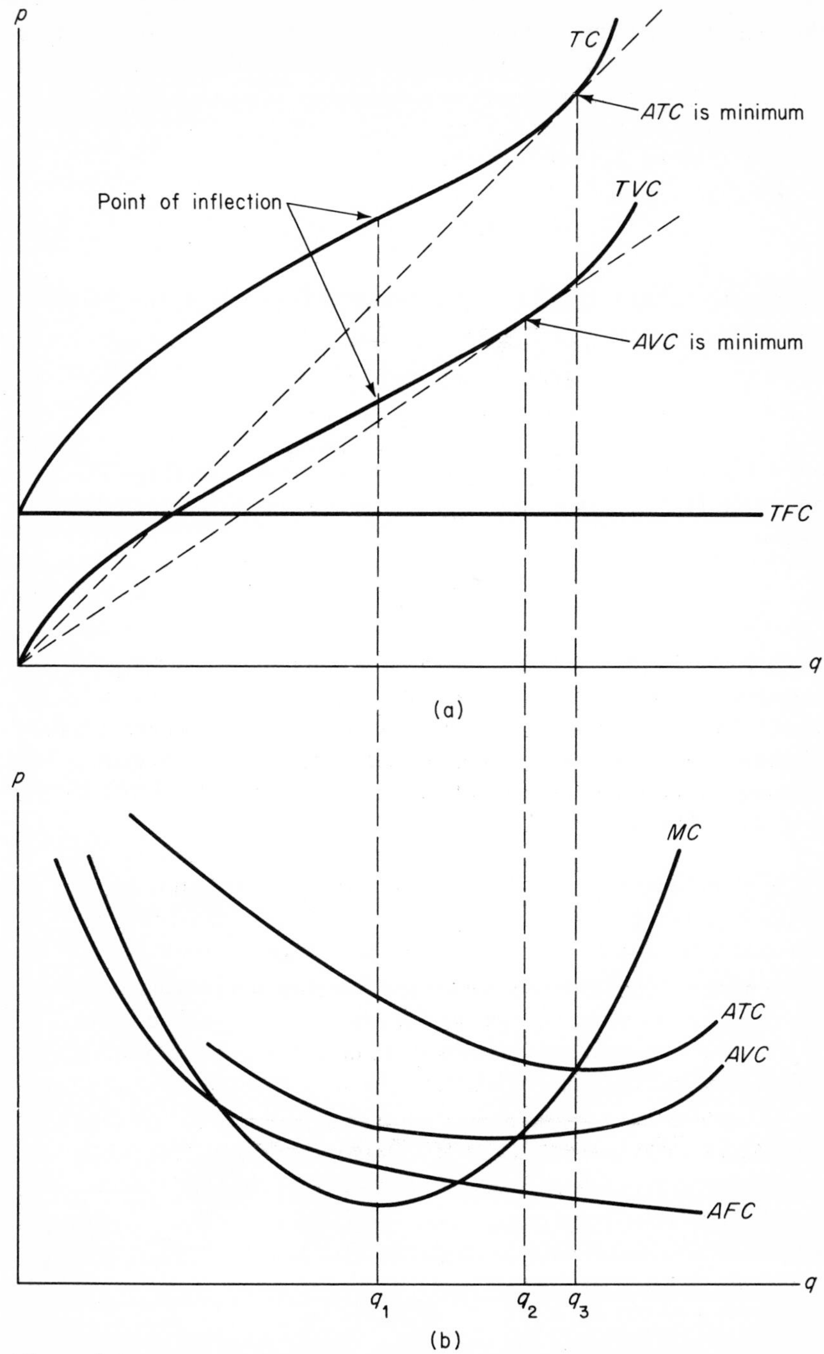
p
TC
ATC is minimum
Point of inflection
TVC
AVC is minimum
TFC
q
(a)
p
MC
ATC
AVC
AFC
q
q_1
q_2
q_3
(b)

Figure 3.6

shows that marginal cost and average variable cost are equal when average variable cost is a minimum. Similarly, marginal cost and average total cost are equal when average total cost is a minimum. This relationship can be demonstrated to exist by determining the minimum value of the *AVC* curve and the *ATC* curve. For *ATC* we observe that

$$ATC = \frac{TC}{q}.$$

Using the derivative formula for a quotient and the fact that the derivative of *TC* is *MC*, we obtain

$$\frac{d(ATC)}{dq} = \frac{q(MC) - TC}{q^2} = 0.$$

Multiplying both sides of the equation by q^2 gives $q(MC) - TC = 0$. Transposing *TC* to the right side of the equal sign and dividing the equation by q give

$$MC = \frac{TC}{q^*}, \tag{3.18}$$

This shows that average total costs are at a minimum when marginal cost equals average total cost. Marginal cost equals average total cost at the critical value of q. The same analysis shows that average variable costs are a minimum when marginal cost and average variable costs are equal.

The relationships among the cost curves can be summarized with the aid of Fig. 3.6 as follows:

1. The point of inflection of the total cost curve q_1 corresponds to the minimum point of the marginal cost curve. This occurs when the second derivative of total cost with respect to quantity is 0.
2. The minimum point on the average variable cost curve occurs when marginal cost equals average variable cost and marginal costs are increasing. This point, q_2, is the point of tangency of a line between the origin and the total variable cost curve.
3. Average total cost is a minimum when marginal cost and average total cost are equal and marginal costs are increasing. This point, q_3, is the point of tangency of a line between the origin and the total cost curve.

Example. Total costs of production for the electronic module manufactured by Hess Electronics is (in thousands of dollars):

$$TC = 0.04q^3 - 0.30q^2 + 2q + 1.$$

Determine MC, AC, AVC, AFC, and TVC curves, plot these curves, and show the relationships between the curves.

(i) $MC = \frac{d(TC)}{dq} = 0.12q^2 - 0.60q + 2.$

MC is a minimum when $\frac{d^2(TC)}{dq^2} = 0$. Thus,

$$\frac{d(MC)}{dq} = 0.24q - 0.60 = 0,$$
$$q_1 = 2.5.$$

(ii) $TVC = 0.04q^3 - 0.30q^2 + 2q.$

(iii) $AVC = \frac{TVC}{q} = 0.04q^2 - 0.30q + 2.$

AVC is a minimum when $\frac{d(AVC)}{dq} = 0$ or, alternatively and equivalently, when $MC = AVC$.

$$\frac{d(AVC)}{dq} = 0.08q - 0.30 = 0,$$
$$q_2 = 3.75.$$

(iv) $AC = \frac{TC}{q} = 0.04q^2 - 0.30q + 2 + \frac{1}{q}\cdot$

AC is a minimum when $\frac{d(AC)}{dq} = 0$ or, alternatively and equivalently, when $MC = AC$.

$$\frac{d(AC)}{dq} = 0.08q - 0.30 - \frac{1}{q^2} = 0,$$
$$0.08q^3 - 0.30q^2 - 1 = 0.$$

By evaluating this equation for alternative values of q, we find that $q_3 = 4.4$ is the solution.

(v) $AFC = TFC/q = 1/q.$

These curves are plotted in Fig. 3.7.

The total cost function was provided in the Hess Electronics Company illustration. The analyst will very seldom be so fortunate as to have this function readily available. The more common case will be when the cost function must be determined from accounting data. The following example illustrates a cost function that is linear in the feasible range of production. The method of fitting nonlinear cost functions to a set of data points is explained in Appendix A.

Example. The Emery Company's accounting records indicate that fixed costs on their plastics line of products are \$17,500 per month. The variable

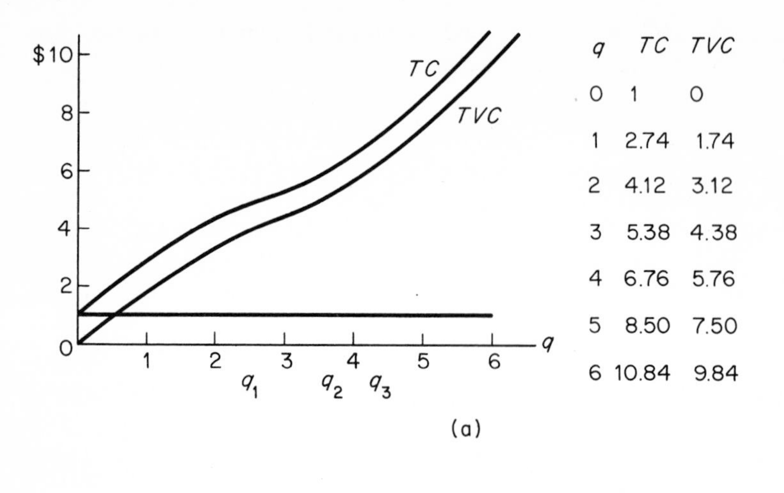

q	TC	TVC
0	1	0
1	2.74	1.74
2	4.12	3.12
3	5.38	4.38
4	6.76	5.76
5	8.50	7.50
6	10.84	9.84

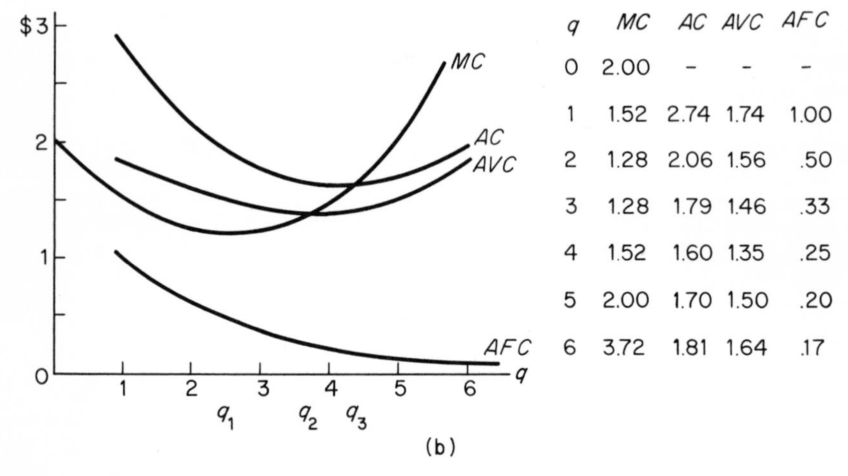

q	MC	AC	AVC	AFC
0	2.00	-	-	-
1	1.52	2.74	1.74	1.00
2	1.28	2.06	1.56	.50
3	1.28	1.79	1.46	.33
4	1.52	1.60	1.35	.25
5	2.00	1.70	1.50	.20
6	3.72	1.81	1.64	.17

Figure 3.7.

costs are $0.60 per dollar of sales. Determine the cost curve for Emery's plastics line.

Since total costs are composed of fixed and variable costs, the total cost per month is

$$TC = 17{,}500 + 0.60X,$$

where X represents the sales volume in dollars.

3.1.3 PROFITS

We are now in a position to combine the revenue and cost functions to obtain an expression for profits. Profits equal total revenue less total cost.

Expressed as a function of the quantity produced and sold, profits are

$$P = TR - TC. \tag{3.19}$$

The quantity that should be produced to obtain maximum profits can be determined by equating the derivative of the profit function with 0. Thus,

$$\frac{d(P)}{dq} = \frac{d(TR)}{dq} - \frac{d(TC)}{dq} = 0. \tag{3.20}$$

The derivative of total revenue with respect to the variable quantity is marginal revenue [formula (3.4)], and the derivative of total cost with respect to quantity is marginal cost [formula (3.13)]. Profits are, therefore, a maximum for the quantity of output for which marginal revenue equals marginal cost.

$$MR = MC. \tag{3.21}$$

The second-order requirement for a maximum is that the second derivative evaluated at the optimum level of output be negative. Thus, for maximum profits, it is also necessary that the second derivative of total revenue minus the second derivative of total cost be negative. That is,

$$\frac{d^2(P)}{dq^2} = \frac{d^2(TR)}{dq^2} - \frac{d^2(TC)}{dq^2} < 0. \tag{3.22}$$

Example. Determine the optimum quantity and the corresponding price for maximum profits for the Hess Electronics Company.

$$P = TR - TC$$

$$P = (3.50q - 0.50q^2) - (0.04q^3 - 0.30q^2 + 2q + 1)$$

$$\frac{d(P)}{dq} = 3.50 - 1.00q - 0.12q^2 + 0.60q - 2 = 0$$

$$-0.12q^2 - 0.40q + 1.50 = 0.$$

Solving for q by using the quadratic formula gives the positive root, $q^* = 2.25$. The second derivative is

$$\frac{d^2(P)}{dq^2} = -0.24q - 0.40,$$

and the second derivative evaluated at $q^* = 2.25$ is less than 0. Therefore, $q^* = 2.25$ is the value of q that leads to maximum profits. Substituting $q^* = 2.25$ in the profit function gives $P = \$910$. From the demand function,

$$p = 3.50 - 0.50q.$$

The price which leads to maximum profits is $p = \$2.38$.

The total revenue, total cost, marginal revenue, marginal cost, and profit functions for Hess Electronics are shown in Fig. 3.8.

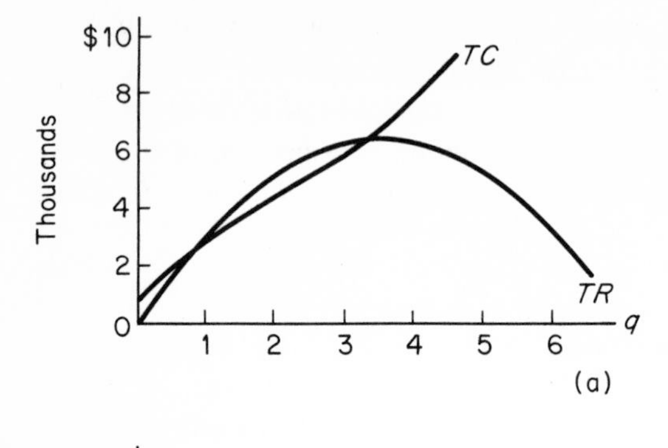

	Thousands		
q	TR	TC	Profit
0	0	1.00	−1.00
1	3.00	2.74	.26
2	5.00	4.12	.88
3	6.00	5.38	.62
4	6.00	6.76	−.76
5	5.00	8.50	−3.50
6	3.00	10.84	−7.84

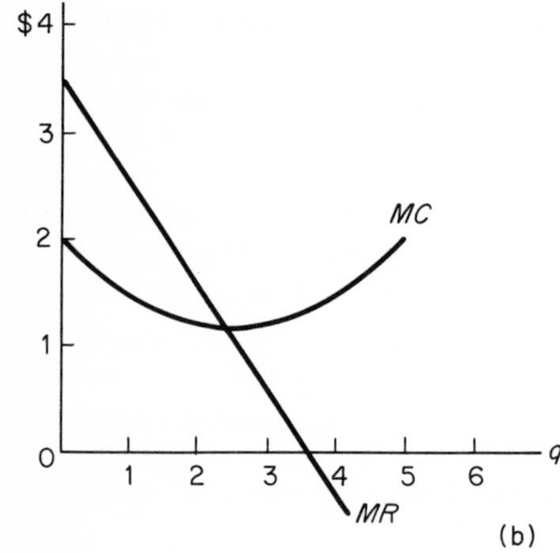

q	MR	MC
0	3.50	2.00
1	2.50	1.52
2	1.50	1.28
3	0.50	1.28
4	−0.50	1.52
5	−1.50	2.00
6	−2.50	3.72

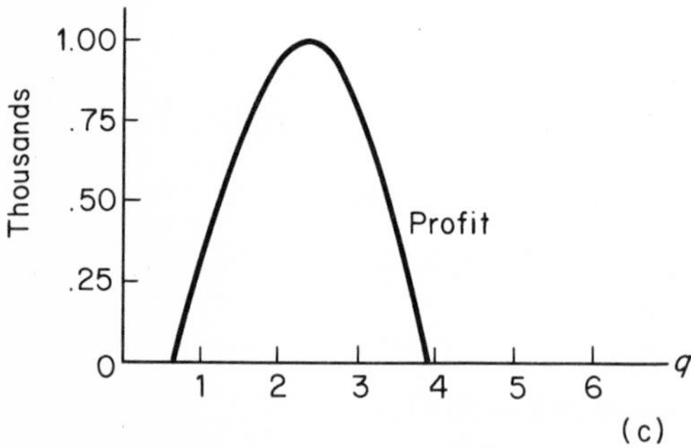

Figure 3.8

3.2 production models

Much effort has been expended in the scientific analysis of production systems. This area of study is termed *production and inventory theory*. Our purpose is to introduce several of the inventory and production models that are employed in the analysis and control of production systems.

Production models are concerned with determining the optimum lot size for production of items for which continual production is not required. For example, machine shops often produce one product or component for a period of time and, after accumulating sufficient stock, change over to production of a different product or component. Similarly, textile mills produce certain quantities of a specific color and texture of material. Orders for this material are then filled from the stock of inventory. As another example, toy manufacturers use the same basic equipment to produce many different toys. A decision must be made, therefore, as to the production quantity size of a given toy. After completing the production run, the equipment is used to manufacture another toy, and orders for the first toy are filled from inventory.

Inventory models are a form of production models. The major difference between the two classifications of models is in the method of obtaining inventory. In the production model, inventory represents the excess of the production rate over the usage rate. In the inventory model, inventory is purchased from the manufacturer (or wholesaler). As an example of an inventory model, book publishing companies order from printing companies quantities of books that are stored in inventory, and orders are filled from this inventory. In a similar fashion, retailers order quantities of merchandise and fill customers' orders from this inventory of goods. In both examples, the inventory model is used for guidance in determining the appropriate order quantity and the proper timing of the order.

The importance of the order quantity or production run quantity stems from the economic value of inventory. Inventory is defined as a stock or supply of idle resources that has economic value. This usually refers to physical goods and materials. It can, however, also be interpreted to include control of any idle resources, including labor, cash, capital goods, and storage space.

The inventory or production model considers the cost of inventory. There are three costs that must be incorporated into the model. These are (1) the cost of carrying inventory, (2) the cost of ordering the inventory or the setup cost associated with a production run, and (3) the cost of the goods produced or ordered.

The cost of carrying inventory includes such items as warehouse expense, insurance, and the cost of the funds that are tied up in inventory. These costs are directly proportional to the quantity of inventory maintained by a firm. The larger the average level of inventory, the greater will be the expense required for storing, insuring, and financing the inventory. If cost of carrying inventory were the only cost considered in the model, inventory costs would be minimized by maintaining no inventory. It is impractical, however, for many firms to maintain no inventory. Sales are lost to competitors who can offer immediate delivery. Consequently, we shall consider only that model in which inventory is maintained.

The cost of ordering the inventory or the setup cost associated with a production run includes costs such as preparing orders or down time resulting from the setup. These costs are a function of the number of orders or setups per period of time. For example, placing orders daily would be more expensive than placing the order only once a year. Thus, the cost of ordering the inventory, or the setup cost associated with a production run, also varies with the order quantity or production run size. If the orders are placed frequently, with resulting small order quantities per order, the cost of ordering is relatively high. Conversely, if the orders are placed infrequently, with resulting large order quantities per order, the cost of ordering is relatively low. Consequently, the cost of ordering is inversely related to the order quantity or production run.

The third cost is that of purchasing or manufacturing the inventory. If we assume a given demand and, for the moment, ignore any possible volume discounts or economies of scale, the total cost of the inventory for the entire period is unrelated to the specific order quantity or production run sizes. In the simpler models, the total cost of inventory is assumed as constant. In more complicated models, factors such as volume discounts and economies of scale resulting from long production runs are included.

In developing the production and inventory models, we assume that demand is known and is uniformly distributed throughout the period. This type of model is classified as a deterministic inventory model.† An alternative model would be a stochastic model in which demand is described by a probability distribution. For the deterministic model, two operating decisions (control variables) are required:

1. The size of the order, i.e., what quantity of goods should be ordered or produced.
2. When to place the order or begin production. We shall assume that this decision is based upon the number of units remaining in inventory.

We first consider the problem of order size. The objective is to determine production schedules so as to minimize total costs. We shall use the following notation:

a = cost of holding one unit of inventory for one time period.

c = setup cost incurred when a new production run is started or inventory is purchased.

---

†The assumption that demand is known simplifies the model. In a good number of practical situations, demand is contractual or is fairly uniform during the period, and this assumption is a reasonable approximation to reality.

k = production rate, in units per time period.
r = rate of demand in units per time period.
q = quantity produced or ordered at each setup.

We shall discuss two basic models in this chapter and two more complicated models in Chapter 5.

3.2.1 ECONOMIC ORDER QUANTITY MODEL

The objective of the economic order quantity model is to determine the order size such that the cost of ordering and carrying inventory is a minimum. It is assumed that shortages are not permitted and that inventory is replenished a fixed number of days following ordering of the inventory.

As an illustration of the model, assume that we have contracted to supply 10,000 assemblies per year. A component of this assembly is ordered from a supplier. The order costs associated with placing an order are \$300. The cost of holding the component as inventory (insurance, warehouse space, cost of capital, etc.) is \$10 per year. The lead time for inventory delivery is five days. Our problem is to determine the order quantity that minimizes total cost. The model is illustrated in Fig. 3.9.

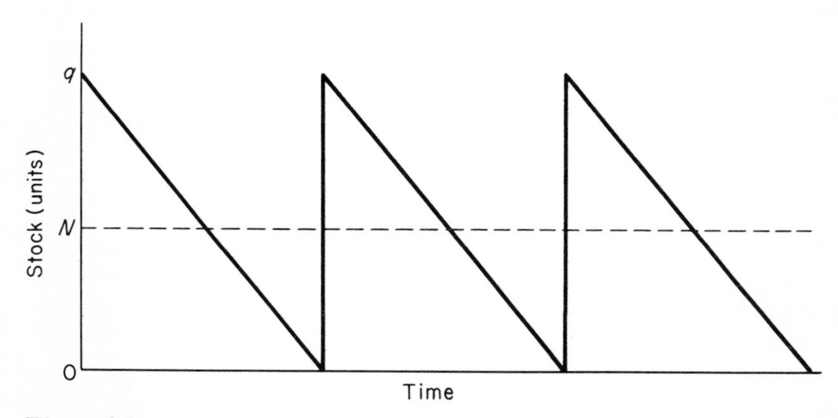

Figure 3.9

The order quantity is represented in Fig. 3.9 by q. The letter N represents the inventory level at which an order must be placed to assure delivery before the inventory is exhausted. The objective is to determine the values of q and N such that the cost of maintaining the inventory is a minimum.

The cost of inventory is composed of an order cost, an inventory carrying cost, and the cost of the inventory:

$$TC = f(q) = C_o + C_c + C_i. \tag{3.23}$$

Consider first the total order cost per period. The number of units demanded during the period is r and the order quantity is q. Thus, r/q represents the number of orders during the period, and the total cost of ordering the inventory is

$$C_o = \frac{rc}{q}. \tag{3.24}$$

It can be seen that the order cost decreases as the order size increases.

The cost of carrying inventory during the period is given by

$$C_c = \frac{qa}{2}, \tag{3.25}$$

where $q/2$ represents the average inventory level. The cost of carrying inventory increases as the order size increases.

The cost of the inventory based upon a given level of demand is constant and is represented by C_i. The total inventory cost is thus

$$TC = \frac{rc}{q} + \frac{qa}{2} + C_i. \tag{3.26}$$

To minimize total cost, we determine the value of q such that $\dfrac{d(TC)}{dq} = 0$. Thus,

$$\frac{d(TC)}{dq} = \frac{-rc}{q^2} + \frac{a}{2} = 0$$

and

$$q^* = \sqrt{\frac{2rc}{a}}. \tag{3.27}$$

q^* is termed the economic order quantity.

The fact that q^* is a minimum can be verified by the second derivative. Since the second derivative of TC,

$$\frac{d^2(TC)}{dq} = \frac{2rc}{q^3},$$

is positive for all positive values of q, we conclude that q^* gives the order quantity that minimizes total inventory cost.

The order point is determined by the rate of inventory usage and the lead time. Let U represent the inventory consumption rate during lead time, and let L represent the lead time. The inventory consumed is given by

$$N = L \cdot U.$$

Example. For the illustrative problem above, determine the economic order quantity and the order point.

In this problem, $r = 10{,}000$, $a = \$10$, and $c = \$300$. Therefore,

$$q = \sqrt{\frac{2(10{,}000)(300)}{10}} = \sqrt{600{,}000} = 776 \text{ units.}$$

We should order 776 components per order.

Assuming that inventory usage is constant and that there are 250 working days per year, we have

$$U = \frac{10{,}000}{250} = 40 \text{ units per day.}$$

If the lead time is five working days, then the order should be placed when the inventory level falls to

$$Nr = L \cdot U = 5 \cdot 40 = 200 \text{ units.}$$

The basic economic order quantity model can be applied directly only if all assumptions underlying the model are reasonably satisfied. One assumption that often is not met is that of constant total cost of the inventory. Quite often volume discounts are allowed, thus encouraging large purchases. These volume discounts must be considered when one is determining the economic order quantity.

The most common form of volume discount is the price break. For example, a manufacturer might offer a price of \$20 per unit on orders up to 200 units. For orders exceeding 200 units, the manufacturer gives a 5 percent discount. A price break of \$1 per unit is thus offered on purchases of 201 units or more.

This price break or volume discount can be considered in the economic order quantity model. The procedure involves computing the economic order quantity for the several prices. In our example, we would determine the economic order quantity for a price per unit of \$20 and the economic order quantity for a price per unit of \$19. If the economic order quantity based upon a price per unit of \$19 is more than 200 units, the problem is solved. The quantity ordered is merely that quantity calculated from the E.O.Q. formula. It often happens, however, that the economic order quantity based upon the \$19 price is less than 200 units. Orders of 200 units or less will not be accepted at a price of \$19. Using formula (3.26), we compare the total cost if we assume orders of 201 units and a price of \$19 per unit with the total cost if we assume the \$20 per unit price and the economic order quantity calculated using formula (3.27). The quantity ordered would be that which results in minimum total cost. The calculations are carried out in the following example.

Example. The regional distributor for the southern region of Universal Electric Appliance Company expects to sell 100 blenders per month during the coming year. For orders of 200 blenders or less the price per blender is \$20. On orders exceeding 200 blenders, the price is reduced to \$19. The cost of storing a blender for one year including possible obsolescence is 20 percent of the price. The ordering cost is \$50 per order. Determine the order quantity.

We shall first determine the economic order quantities, assuming storage costs of \$4.00 per unit for units purchased for \$20 and \$3.80 per unit for units purchased for \$19.

$$q_1 = \sqrt{\frac{2(1200)(50)}{4.00}} = 173 \text{ units}$$

$$q_2 = \sqrt{\frac{2(1200)(50)}{3.80}} = 178 \text{ units.}$$

It is possible to order quantities of 173 units at the price of \$20 per unit. To receive the price of \$19 per unit, we must order quantities of 201 units. The order quantity that results in minimum total cost can be determined from (3.25).

$$\begin{aligned} TC_1 &= \frac{1200(50)}{173} + \frac{1}{2}(173)(4.00) + 1200(20) \\ &= 347.00 + 346.00 + 24{,}000 = \$24{,}693 \\ TC_2 &= \frac{1200(50)}{201} + \frac{1}{2}(201)(3.80) + 1200(19) \\ &= 299.00 + 381.00 + 22{,}800 = \$23{,}480. \end{aligned}$$

Order quantities of 173 units results in total inventory cost of \$24,693 and order quantities of 201 units results in total inventory cost of \$23,480. A saving of \$1213 is the result of the order quantity of 201 units. The appropriate order quantity is thus 201 units.

3.2.2 PRODUCTION LOT SIZE MODEL

The objective of the production lot size model is to determine the quantity of product to be manufactured during a production run. The costs incorporated in the model are the cost of setup, the cost of carrying the inventory, and the cost of the inventory. The production lot size is that quantity which results in the minimum total cost. This model differs from the economic order quantity model in that it was assumed in the economic order quantity model that the quantity ordered is, upon arrival, immediately added to inventory. In the production lot size model it is assumed that the product being produced is added to inventory over a period of time.

As an example, consider the manufacturing situation in which a component is manufactured and stored in inventory. The annual demand is

$r = 2000$ units, and the production rate is $k = 12{,}000$ units. Holding costs are $a = \$10$ per unit per year and the setup cost is $c = \$250$ per setup. Determine the production lot size q.

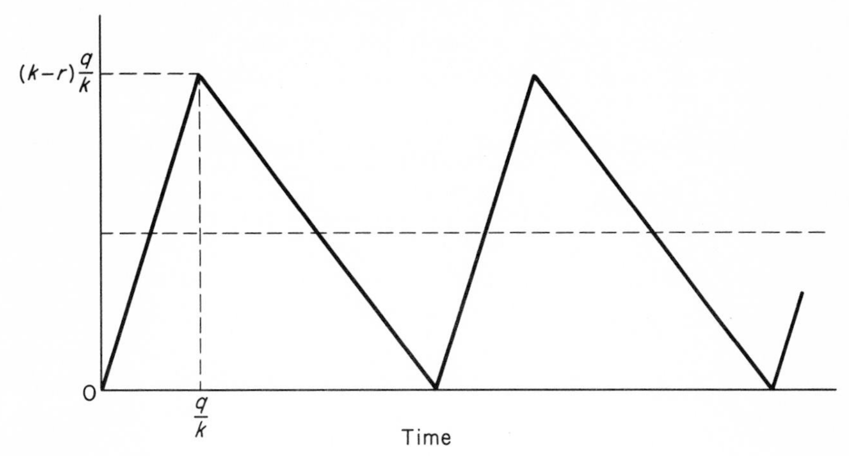

Figure 3.10

Total cost per period is composed of setup cost, carrying cost, and the cost of the inventory:

$$TC = f(q) = C_s + C_c + C_i. \qquad (3.28)$$

Since there are r/q setups per period, each costing c, the total setup cost per period is

$$C_s = \frac{rc}{q}. \qquad (3.29)$$

The total carrying cost per period is found by multiplying the average number of units of inventory by the cost per period of carrying one unit of inventory. The term q/k gives the proportion of the period required to manufacture q units. Since units are being used concurrently while they are being produced, the net addition to inventory is at the rate of $(k - r)$. The maximum inventory level is thus $(k - r)q/k$. The average inventory is one-half the maximum level of inventory. The total cost of carrying the inventory is

$$C_c = \frac{q(k - r)a}{2k}. \qquad (3.30)$$

The total cost per period can now be expressed in terms of the production lot size as

$$TC = \frac{rc}{q} + \frac{q(k - r)a}{2k} + C_i. \qquad (3.31)$$

The minimum cost is found by equating the derivative of the total cost function to 0 and solving for q.

$$\frac{d(TC)}{dq} = \frac{-rc}{q^2} + \frac{a(k - r)}{2k} + 0 = 0.$$

$$q^* = \sqrt{\frac{2krc}{a(k - r)}}. \tag{3.32}$$

Since the second derivative of TC, $\frac{d^2(TC)}{dq^2} = \frac{2rc}{q^3}$, is positive for all positive values of q, we conclude that q^* gives the production lot size that minimizes total inventory cost.

Example: For the illustrative problem, determine the production lot size. In this problem, $r = 2000$; $k = 12{,}000$; $a = \$10$; $c = \$250$.

$$q^* = \sqrt{\frac{2(12{,}000)(2000)(250)}{10(12{,}000 - 2000)}} = \sqrt{120{,}000}$$

$$q^* = 346.$$

problems

1. The demand function for a particular product is $p = 5.00 - 0.75q$. Determine the total revenue, marginal revenue, average revenue, and elasticity of demand functions. Find the value of q that results in maximum revenue.
2. The demand function for a commodity is $p = 1000 - 50q - q^2$ for $0 < q < 25$. Determine the total revenue, marginal revenue, average revenue, and elasticity of demand functions. Find the value of q that results in maximum total revenue, and determine the total revenue, marginal revenue, average revenue, and elasticity of demand for this value of q.
3. The demand function for a product is $p = 2500 - 60q - 0.5q^2$ for $0 < q < 60$. Determine the total revenue, marginal revenue, average revenue, and elasticity of demand functions. Find the value of q that results in maximum revenue, and determine the total revenue, marginal revenue, average revenue, and elasticity for this value of q.
4. The demand function for a product is $p = 10 - 0.25q - 0.05q^2$ for $0 < q < 10$. Determine the total revenue, marginal revenue, average revenue, and elasticity of demand functions. Find the value of q that results in maximum revenue, and determine the total

revenue, marginal revenue, average revenue, and elasticity for this value of q.

5. The ABC Bus Tour Company conducts tours of San Diego. At a price of \$5.00 per person, ABC experienced an average demand of 1000 customers per week. Following a price reduction to \$4.00 per person, the average demand increased to 1200 customers per week. Assuming a linear demand function, determine the total revenue, marginal revenue, average revenue, and elasticity of demand functions. Find the tour fare and quantity that result in maximum revenue, and determine the total revenue, marginal revenue, average revenue, and elasticity of demand for this value of q.
6. The Santa Catalina Motor Excursion Company offers round trip boat transportation between Santa Catalina Island and the Port of Los Angeles. They are in competition with one other boat line plus an airline. Because of the competition, the owners of the company have not been able to raise their price. The owners of the company believe, however, that an increase in the fare would result in an increase in total revenue. An analysis of the market indicates that a \$2.00 increase in the present round trip boat fare of \$16.00 would result in a 5 percent decrease in the number of passengers. Using the concept of elasticity of demand, determine if the price increase should be made.
7. The Santa Catalina Motor Excursion Company of the preceding problem was able to convince the other boat line that a \$2.00 increase would lead to increased revenue. However, the airline could not raise its fare without governmental approval. If the fare charged by the airline remains unchanged, it is estimated that the proposed \$2.00 increase in boat fare would lead to a 15 percent decrease in the number of passengers carried by the boat lines. Should the fare increase be made?
8. The cost function for a particular commodity is a linear function of output for $q \leq 1000$ units. If fixed costs are \$1500 and the variable cost of manufacturing is \$2.50 per unit, determine total cost, marginal cost, and average cost functions. What value of q gives minimum marginal and average cost?
9. The total cost function for a product is described by the function $TC = 1000 + 20q - q^2 + 0.05q^3$ for $0 \leq q \leq 30$. Determine the marginal cost, average variable cost, and average fixed cost functions. Determine the value of q for which marginal cost is a minimum. Determine also the values of q for which average cost and average variable cost are minimum. Verify the fact that marginal cost equals average cost when average cost is a minimum and that

marginal cost equals average variable cost when average variable cost is a minimum. For what value of q is average fixed cost a minimum?

10. The total cost function for a product is described by the function $TC = 4500 + 1500q - 1.5q^2 + 0.02q^3$. Determine the marginal cost, average cost, average variable cost, and average fixed cost functions. Determine the value of q for which marginal cost is a minimum. Determine also the values of q for which average cost and average variable cost are minimum. Verify the fact that marginal cost equals average cost when average cost is a minimum and that marginal cost equals average variable cost when average variable cost is a minimum.
11. The total revenue function for a particular product is $TR = 600q - 0.5q^2$. The total cost for the product is $TC = 1500 + 150q - 4q^2 + 0.5q^3$. Determine the profit function and the value of q for which profits are maximum. Is this also the value of q for which revenue is a maximum or for which average costs are a minimum?
12. The total revenue function for a product is $TR = q(400 - q)$, and the total cost for the product is $TC = 1000 + 100q - 5q^2 + q^3$. Determine marginal revenue, marginal cost, the output for which profit is a maximum, and the profit at this level of output.
13. The demand for a commodity is $p = 200 - q$. The cost of producing the commodity is composed of a fixed cost of \$5000 and a variable cost of \$0.50 per unit. Determine the profit function for the commodity and the quantity of output that leads to maximum profit.
14. A tour charges groups of individuals \$100 per person less \$1 per person for each individual in the group over 25, that is, for example, 26 members of a group would pay \$99 each, and the total revenue to the tour organization would be \$99(26). The cost of organizing and conducting the tour is \$1000 plus \$40 per person. What group size leads to maximum profit, and what is this profit?
15. The Western Petroleum Company purchases auto accessories from independent manufacturers and sells these products under the Western Petroleum label through their chain of automobile service stations. Batteries are purchased from the Universal Power Company. The yearly demand for batteries is estimated as 25,000 units. The per year cost of holding the battery, including a charge for possible obsolescence, is \$5. The cost of placing an order with Universal Power is \$200. Determine the economic order quantity.
16. Western Petroleum expects to sell 100,000 tires during the year. The cost of holding a tire in inventory for one year is \$4, and the

cost of ordering the tires is $150. Determine the economic order quantity.

17. Universal Power uses the same equipment to manufacture both 6-volt and 12-volt batteries. The yearly demand for 6-volt batteries is 75,000 units, and the demand for 12-volt batteries is 225,000 units. Production capacity is such that 500,000 batteries can be manufactured each year. The cost of changing from the production of one battery to the other is $500; i.e., setup cost is $500. The cost of holding a battery is $6 per year. Determine the production lot sizes for both batteries.
18. The Haynes Music Corporation has leased 300 acres of beach front between San Diego and Los Angeles for a rock music festival. The rock music groups have been engaged, and all other necessary arrangements have been made. Consequently, all major costs have been incurred. Haynes is uncertain what admission price should be charged for the festival. Assuming the objective of maximizing total revenue, determine the appropriate price. The following estimates of demand are available.

Price	*Number of customers*
$15	50,000
10	150,000
5	200,000

4

calculus of multivariate and transcendental functions

This chapter continues the discussion of the concepts and techniques that underlie the application of differential calculus in business and economics. We shall discuss multivariate functions, transcendental functions, and the method of Lagrangian multipliers. Applications of these concepts are given in this chapter and in Chapter 5.

4.1 multivariate functions

In the discussion and problems of Chapters 1 and 2, we have considered only functions of one independent variable, expressed as $y = f(x)$. In many important problems, however, the single dependent variable y is related to more than one independent variable. That is,

$$y = f(x_1, x_2, x_3). \tag{4.1}$$

The functional relationship specified in (4.1) is termed a *multivariate function.* x_1, x_2, and x_3 are the independent variables, and y is the single dependent variable. For each combination of values of x_1, x_2, and x_3 that are permitted by the structure of the problem, the dependent variable assumes a value that is specified by the functional relationship.

Example. Given that $f(x_1, x_2) = 2x_1 - 3x_2 + 6$, calculate $f(x_1, x_2)$ for the following data points.

$$\begin{aligned} f(4, 2) &= 2(4) - 3(2) + 6 = 8 \\ f(3, 7) &= 2(3) - 3(7) + 6 = -9 \\ f(7, 3) &= 2(7) - 3(3) + 6 = 11 \\ f(7, 4) &= 2(7) - 3(4) + 6 = 8. \end{aligned}$$

The example illustrates that for each combination of values of x_1 and x_2, there is only one value of $f(x_1, x_2)$. This requirement does not imply that the function evaluated for different combinations of variables cannot have the same value. In the above example,

$$f(4, 2) = 8 \quad \text{and} \quad f(7, 4) = 8.$$

It is not, however, possible for $f(4, 2)$ to equal any value other than 8 for the function considered in the example. For each combination of values of the independent variables, there is one and only one value of the dependent variable. If the structure of the problem requires more than one value of the dependent variable for a specific combination of values of the independent variables, the problem requires the mathematics of multivariate relations. Mathematics of multivariate relations are not considered in this text.

Example. If $f(x_1, x_2, x_3) = x_1^2 + 2x_1x_2 + x_2^2 + x_2x_3$, calculate the value of $f(x_1, x_2, x_3)$ for the following data points:

$$\begin{aligned} f(3, 4, 5) &= (3)^2 + 2(3)4 + (4)^2 + 4(5) = 69 \\ f(2, -1, 3) &= (2)^2 + 2(2)(-1) + (-1)^2 + (-1)3 = -2 \\ f(1, 0, 1) &= (1)^2 + 2(1)(0) + (0)^2 + 0(1) = 1 \\ f(2, 4, 6) &= (2)^2 + 2(2)4 + (4)^2 + 4(6) = 60. \end{aligned}$$

This is an example of a function in which the single dependent variable, represented by $f(x_1, x_2, x_3)$, is related to three independent variables. The value of the dependent variable is determined by evaluating the function for the specified values of the independent variables. For each combination of values of the independent variables, there exists only one value of the dependent variable.

The following three examples illustrate business and economic applications of multivariate functions.

Example. Assume that sales S is a function of both price P and advertising A. If

$$S = 500(10 - P)^2A^{1/2},$$

determine S for $P = \$4$ and $A = \$10{,}000$.

$$\begin{aligned} S &= 500(6)^2(10{,}000)^{1/2} \\ S &= 500(36)(100) = \$1{,}800{,}000. \end{aligned}$$

Example. Assume that labor and capital can, within a limited range, be substituted for each other. If output Y is a function of labor L and capital K as described below, calculate output for $L = 1000$ and $K = 40{,}000$.

$$\begin{aligned} Y &= 10(L)^{2/3} + 0.50(K)^{1/2} \\ Y &= 10(1000)^{2/3} + 0.50(40{,}000)^{1/2} \\ Y &= 10(100) + 0.50(200) = 1100. \end{aligned}$$

Example. Assume that the Neshay Candy Company makes a net profit of \$0.05 per almond bar and \$0.07 per chocolate bar. If net profit P is represented as a function of monthly sales in units of the almond bar x_1 and the chocolate bar x_2, determine the appropriate functional relationship and calculate net profit for a month in which $x_1 = 10{,}000$ and $x_2 = 20{,}000$.

$$\begin{aligned} P &= 0.05x_1 + 0.07x_2 \\ P &= 0.05(10{,}000) + 0.07(20{,}000) \\ P &= \$1900. \end{aligned}$$

Multivariate functions can be established through data points by using the technique described in Appendix A. This technique involves substituting the appropriate data points into the specified general form of the multivariate function and solving the resulting equations simultaneously for the parameters of the function. The technique is illustrated by the following example.

Example. Research has demonstrated that the output of a firm Y is related to labor L and capital K. The general form of the function is

$$Y = aL + bK + c(LK).$$

It has been determined that: $Y = 100{,}000$ when $L = 9.0$ and $K = 4.0$; $Y = 120{,}000$ when $L = 10.0$ and $K = 5.0$; and $Y = 150{,}000$ when $L = 11.5$ and $K = 7.0$. The domains of L and K are $9.0 \leq L \leq 11.5$ and $4.0 \leq K \leq 7.0$. Determine the values of a, b, and c.

The three data points are substituted into the general form of the function to give

$$\begin{aligned} 100{,}000 &= a(9.0) + b(4.0) + c(36.0) \\ 120{,}000 &= a(10.0) + b(5.0) + c(50.0) \\ 150{,}000 &= a(11.5) + b(7.0) + c(80.5). \end{aligned}$$

These equations are solved simultaneously to give $a = -60{,}000$, $b = 304{,}000$, and $c = -16{,}000$. The function is

$$Y = -60{,}000L + 304{,}000K - 16{,}000(LK) \qquad \text{for } 9.0 \leq L \leq 11.5, \quad 4.0 \leq K \leq 7.0.$$

4.1.1 DERIVATIVES OF MULTIVARIATE FUNCTIONS

The concept of the derivative extends directly to multivariate functions. In our discussion of derivatives in Chapter 1, we defined the derivative of a function as the instantaneous rate of change of the function with respect to the independent variable. In multivariate functions, the derivative again gives the instantaneous rate of change of the function with respect to an independent variable. Since, however, there is more than one independent variable in multivariate functions, the derivative of the function must be considered separately for each independent variable. The derivative of the function with respect to one of the independent variables is termed the *partial derivative* of the function.

The formula for the partial derivative of the multivariate function with respect to an independent variable is given by (4.2). If $y = f(x_1, x_2)$, the derivative of y with respect to x_1 is

$$\frac{\delta y}{\delta x_1} = \underset{\Delta x_1 \to 0}{\text{limit}} \frac{f(x_1 + \Delta x_1, x_2) - f(x_1, x_2)}{\Delta x_1}, \tag{4.2}$$

where the symbol $\frac{\delta y}{\delta x_1}$ is read *the partial derivative of y with respect to x_1*. The partial derivative of y with respect to x_2 is expressed as

$$\frac{\delta y}{\delta x_2} = \underset{\Delta x_2 \to 0}{\text{limit}} \frac{f(x_1, x_2 + \Delta x_2) - f(x_1, x_2)}{\Delta x_2}, \tag{4.3}$$

where $\frac{\delta y}{\delta x_2}$ is read *the partial derivative of y with respect to x_2*.

The partial derivative of a multivariate function gives the instantaneous rate of change of the dependent variable with respect to an independent variable. To determine the effect of a change in a single independent variable upon the dependent variable, the remaining independent variables must not be allowed to change in value. It can be seen that if all independent variables were allowed to vary concurrently, it would be impossible to determine the effect of a change in a single variable. Thus, all variables other than the variable under consideration are assumed to be constant.

The partial derivatives of the multivariate function can be determined from the formulas for derivatives of multivariate functions. To illustrate the use of these formulas, we first determine the partial derivatives of $y = 4x_1 + 5x_2$. The partial derivative of y with respect to x_1 is given by formula (4.2).

$$\frac{\delta y}{\delta x_1} = \underset{\Delta x_1 \to 0}{\text{limit}} \frac{f(x_1 + \Delta x_1, x_2) - f(x_1, x_2)}{\Delta x_1}. \tag{4.2}$$

According to formula (4.2), x_1 is replaced by $x_1 + \Delta x_1$. This yields

$$\frac{\delta y}{\delta x_1} = \underset{\Delta x_1 \to 0}{\text{limit}} \left[\frac{4(x_1 + \Delta x_1) + 5x_2 - (4x_1 + 5x_2)}{\Delta x_1}\right]$$

$$\frac{\delta y}{\delta x_1} = \underset{\Delta x_1 \to 0}{\text{limit}} \frac{4\,\Delta x_1}{\Delta x_1} = 4.$$

The partial derivative of y with respect to x_2 is given by formula (4.3).

$$\frac{\delta y}{\delta x_2} = \underset{\Delta x_2 \to 0}{\text{limit}} \frac{f(x_1, x_2 + \Delta x_2) - f(x_1, x_2)}{\Delta x_2}. \tag{4.3}$$

In the example problem, we replace x_2 by $x_2 + \Delta x_2$.

$$\frac{\delta y}{\delta x_2} = \underset{\Delta x_2 \to 0}{\text{limit}} \left[\frac{4x_1 + 5(x_2 + \Delta x_2) - (4x_1 + 5x_2)}{\Delta x_2}\right]$$

$$\frac{\delta y}{\delta x_2} = \underset{\Delta x_2 \to 0}{\text{limit}} \frac{5\,\Delta x_2}{\Delta x_2} = 5.$$

Example. Determine the partial derivatives of $f(x_1, x_2) = 3x_1 + 6x_2$.

$$\frac{\delta f}{\delta x_1} = \underset{\Delta x_1 \to 0}{\text{limit}} \left[\frac{3(x_1 + \Delta x_1) + 6x_2 - (3x_1 + 6x_2)}{\Delta x_1}\right]$$

$$\frac{\delta f}{\delta x_1} = \underset{\Delta x_2 \to 0}{\text{limit}} \frac{3\,\Delta x_1}{\Delta x_1} = 3,$$

$$\frac{\delta f}{\delta x_2} = \underset{\Delta x_2 \to 0}{\text{limit}} \left[\frac{3x_1 + 6(x_2 + \Delta x_2) - (3x_1 + 6x_2)}{\Delta x_2}\right]$$

$$\frac{\delta f}{\delta x_2} = \underset{\Delta x_2 \to 0}{\text{limit}} \frac{6\,\Delta x_2}{\Delta x_2} = 6.$$

For the multivariate function $y = f(x_1, x_2)$, the symbol for the partial derivative of y with respect to x_1 is $\frac{\delta y}{\delta x_1}$, and the symbol for the partial derivative of y with respect to x_2 is $\frac{\delta y}{\delta x_2}$. Alternatively, the partial derivative of the function with respect to x_1 can be expressed as f_{x_1} and the partial derivative of the function with respect to x_2 as f_{x_2}.

The derivatives of nonlinear multivariate functions are determined in the same manner as the derivatives of linear multivariate functions. The partial derivative of the function with respect to any independent variable is found by considering the remaining independent variables as constant. This is illustrated by the following examples.

Example. Determine the partial derivatives of $y = x_1^2 + x_1x_2 + x_2^2$. Application of the formula for the partial derivative of y with respect to x_1 gives

$$\frac{\delta y}{\delta x_1} = \underset{\Delta x_1 \to 0}{\text{limit}} \frac{[(x_1 + \Delta x_1)^2 + (x_1 + \Delta x_1)x_2 + x_2^2] - [x_1^2 + x_1x_2 + x_2^2]}{\Delta x_1}$$

$$\frac{\delta y}{\delta x_1} = \underset{\Delta x_1 \to 0}{\text{limit}} \frac{(x_1^2 + 2x_1\Delta x_1 + \Delta x_1^2 + x_1x_2 + \Delta x_1x_2 + x_2^2 - x_1^2 - x_1x_2 - x_2^2)}{\Delta x_1}$$

$$\frac{\delta y}{\delta x_1} = \underset{\Delta x_1 \to 0}{\text{limit}} \frac{(2x_1\,\Delta x_1 + \Delta x_1^2 + \Delta x_1 x_2)}{\Delta x_1} = \underset{\Delta x_1 \to 0}{\text{limit}} (2x_1 + \Delta x_1 + x_2)$$

$$\frac{\delta y}{\delta x_1} = 2x_1 + x_2.$$

The partial derivative of y with respect to x_2 is

$$\frac{\delta y}{\delta x_2} = \underset{\Delta x_2 \to 0}{\text{limit}} \frac{[x_1^2 + x_1(x_2 + \Delta x_2) + (x_2 + \Delta x_2)^2] - [x_1^2 + x_1 x_2 + x_2^2]}{\Delta x_2}$$

$$\frac{\delta y}{\delta x_2} = \underset{\Delta x_2 \to 0}{\text{limit}} \frac{(x_1^2 + x_1 x_2 + x_1 \Delta x_2 + x_2^2 + 2x_2 \Delta x_2 + \Delta x_2^2 - x_1^2 - x_1 x_2 - x_2^2)}{\Delta x_2}$$

$$\frac{\delta y}{\delta x_2} = \underset{\Delta x_2 \to 0}{\text{limit}} \frac{(x_1\,\Delta x_2 + 2x_2\,\Delta x_2 + \Delta x_2^2)}{\Delta x_2} = \underset{\Delta x_2 \to 0}{\text{limit}} (x_1 + 2x_2 + \Delta x_2)$$

$$\frac{\delta y}{\delta x_2} = x_1 + 2x_2.$$

In each of these examples, x_2 is considered a constant when one is determining the partial derivative of y with respect to x_1. Similarly, x_1 is considered a constant when one is determining the partial derivative of y with respect to x_2. Since we are interested in determining the partial derivative of y with respect to only one variable at a time and are treating the other variables as constants, the multivariate function can be considered as a function of only one independent variable for purposes of differentiation. The other variables are treated as constants in applying the rules for finding derivatives.

Consider again the previous examples. If $y = 4x_1 + 5x_2$, and x_2 is treated as a constant, then $\delta y/\delta x_1 = 4$. Similarly, if x_1 is assumed constant, $\delta y/\delta x_2 = 5$. In the third example, treating x_2 as a constant and determining the derivative with respect to x_1, we have $\delta y/\delta x_1 = 2x_1 + x_2$. Again, $\delta y/\delta x_2 = x_1 + 2x_2$.

The rules for determining derivatives of multivariate functions are the same as those for finding derivatives of functions with only one independent variable. In applying these rules, remember that all variables other than the variable for which the derivative is taken with respect to are treated as constants.

Example. Determine the partial derivatives of $y = 3x_1^2 + 2(x_1x_2)^3 + x_2^4$. To determine the partial derivative of y with respect to x_1, we treat x_2 as a constant.

$$\frac{\delta y}{\delta x_1} = 6x_1 + 6(x_1x_2)^2(x_2) + 0.$$

Similarly, in determining the partial derivative of y with respect to x_2, we consider x_1 to be constant.

$$\frac{\delta y}{\delta x_2} = 0 + 6(x_1x_2)^2(x_1) + 4x_2^3.$$

The same procedure is applied in the following examples.

Example. $y = (2x_1 + x_2)^5$.

$$\frac{\delta y}{\delta x_1} = 5(2x_1 + x_2)^4(2)$$

$$\frac{\delta y}{\delta x_2} = 5(2x_1 + x_2)^4(1).$$

Example. $y = 3x_1^2 + 2x_1x_2x_3 + 3x_2x_3^3 + x_3^4$.

$$\frac{\delta y}{\delta x_1} = 6x_1 + 2x_2x_3$$

$$\frac{\delta y}{\delta x_2} = 2x_1x_3 + 3x_3^3$$

$$\frac{\delta y}{\delta x_3} = 2x_1x_2 + 9x_2x_3^2 + 4x_3^3.$$

Example. $y = \dfrac{2x_1 - x_2^2}{x_1 + x_2}$.

$$\frac{\delta y}{\delta x_1} = \frac{(x_1 + x_2)(2) - (2x_1 - x_2^2)(1)}{(x_1 + x_2)^2} = \frac{2x_2 + x_2^2}{(x_1 + x_2)^2}$$

$$\frac{\delta y}{\delta x_2} = \frac{(x_1 + x_2)(-2x_2) - (2x_1 - x_2^2)(1)}{(x_1 + x_2)^2}$$

$$\frac{\delta y}{\delta x_2} = \frac{-2x_1x_2 - 2x_1 - x_2^2}{(x_1 + x_2)^2} = \frac{-2x_1x_2 - 2x_1 - x_2^2}{(x_1 + x_2)^2}.$$

4.1.2 INTERPRETATION OF PARTIAL DERIVATIVES

The partial derivative of a function with respect to the variable x_1 describes the change in the function for an incremental change in the value of x_1. If the dependent variable is functionally related to two independent variables, this relationship can be shown by a three-dimensional graph. The partial derivative of y with respect to x_1 gives the slope of the function in a plane parallel to the plane formed by the y and x_1 axes. The position of this plane is x_2 units from the origin. We can express the same concept by the statement that the partial derivative of y with respect to x_1 gives the rate of change of the function for an incremental change in x_1, assuming that x_2 is held constant. Similarly, the partial derivative of y with respect to x_2 gives the slope of the function in a plane parallel to the plane formed by the

y and x_2 axes. The position of the plane is x_1 units from the origin. An equivalent interpretation is that the partial derivative of y with respect to x_2 gives the rate of change of the function for an incremental change in x_2 with x_1 held constant. The partial derivatives of a function of two independent variables are illustrated by the following example.

Example. $f(x_1, x_2) = x_1^2 + 2x_1x_2 + x_2^2$ describes the three-dimensional surface sketched below. Determine the slope of the function in the plane parallel to the f and x_1 axes located $x_2 = 2$ units from the origin. Also determine the slope of the function in the plane parallel to the f and x_2 axes located $x_1 = 1$ unit from the origin.

The function $f(x_1, x_2) = x_1^2 + 2x_1x_2 + x_2^2$ is sketched in Fig. 4.1(a). The function that describes the surface in the plane parallel to the f and x_1 axes that is located $x_2 = 2$ units from the origin is sketched in Fig. 4.1(b). The function is

$$f(x_1, 2) = x_1^2 + 2x_1(2) + (2)^2$$

$$f(x_1, 2) = x_1^2 + 4x_1 + 4.$$

This function is sketched in Fig. 4.1(b). The partial derivative of $f(x_1, 2)$ gives the slope of the function in this plane.

$$\frac{\delta f}{\delta x_1} = 2x_1 + 4.$$

The derivative is sketched in Fig. 4.1(d).

The function that describes the surface in the plane parallel to the f and x_2 axes that is $x_1 = 1$ unit from the origin is plotted in Fig. 4.1(c) and given below:

$$f(1, x_2) = 1 + 2x_2 + x_2^2.$$

The partial derivative of the function gives the slope of the function in this plane. The partial derivative, plotted in Fig. 4.1(e), is

$$\frac{\delta f}{\delta x_2} = 2 + 2x_2.$$

Functions of more than three dimensions cannot be graphed. Consequently, it is meaningless to speak of the slope of a function of three or more independent variables. Instead, we are interested in the change in the value of the function for an incremental change in an independent variable. This is illustrated by the following examples.

Example. Estimated sales s of a product are related to price p, advertising a, and the number of salesmen n. The functional relationship is

$$s = (10{,}000 - 800p)n^{1/2}a^{1/2}.$$

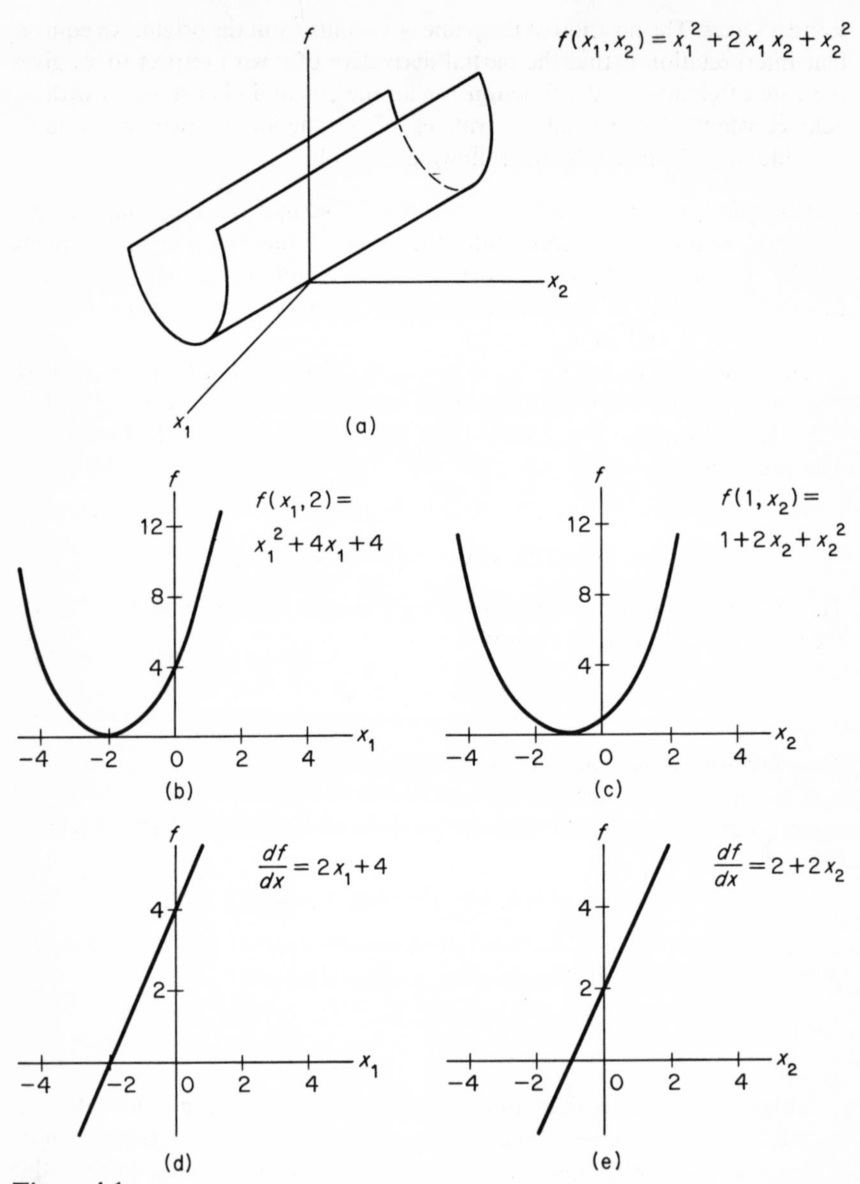

Figure 4.1

If p = \$5 and a = \$10,000, determine the effect of adding additional salesmen. The marginal effect is described by the partial derivative.

$$\frac{\delta s}{\delta n} = \frac{a^{1/2}}{2}(10{,}000 - 800p)n^{-1/2}.$$

If the current number of salesmen is $n = 100$, the marginal effect of an additional salesman is approximated by evaluating the partial with respect to n for $n = 100$, $p = \$5$, and $a = \$10{,}000$.

$$\begin{aligned}\frac{\delta s}{\delta n} &= \tfrac{1}{2}(10{,}000)^{1/2}[10{,}000 - 800(5)](100)^{-1/2} \\ &= \tfrac{1}{2}(100)(6000)\tfrac{1}{10} = \$30{,}000.\end{aligned}$$

The partial derivative represents the marginal change in S for a marginal change in n. We approximate the total change in S for a one-unit change in n by the product of the partial derivative and the unit change in n. Thus,

$$\Delta S \simeq \Delta n \cdot \frac{\delta S}{\delta n},$$

and

$$\Delta S \simeq (1)(\$30{,}000) = \$30{,}000.$$

Similarly, if it were possible to add 0.1 of a salesman's time, the addition to total sales would be \$3000.

Example. The sales of a manufacturing firm have been found to depend upon the amount invested in research and development R, plant and capital equipment P, and marketing M according to the following function:

$$S = 5R^{0.3}P^{0.5}M^{0.4}, \text{ in millions of dollars.}$$

The following amounts are budgeted for investment during the current year; $R = \$12$ million, $P = \$7$ million, and $M = \$10$ million. Determine expected sales, using the predicting function for sales. Determine also the effect of additional expenditures in R, P, and M.

Expected sales can be calculated as follows:†

$$\begin{aligned}S &= 5(12)^{0.3}(7)^{0.5}(10)^{0.4} \\ \log S &= \log 5 + 0.3 \log 12 + 0.5 \log 7 + 0.4 \log 10 \\ \log S &= 0.6990 + 0.3(1.0792) + 0.5(0.8451) + 0.4(1.0000) \\ \log S &= 1.8453 \\ S &= 70.03.\end{aligned}$$

The effect of additional expenditures can be determined from the partial derivatives. The partial derivative of S with respect to R is

$$\frac{\delta S}{\delta R} = 1.5R^{-0.7}P^{0.5}M^{0.4}.$$

An additional expenditure of \$1 in research and development results in a

---

†The student is referred to Appendix B, Sec. B.4, for calculations using logarithms.

marginal increase in sales of $\Delta S = \$1 \cdot \delta S/\delta R$,

$$\frac{\delta S}{\delta R} = 1.5(12)^{-0.7}(7)^{0.5}(10)^{0.4} = \$1.75.$$

Using the same procedure, we find that the marginal increase in sales for a $1 increase in plant and equipment expenditures is $4.96 and the marginal increase in sales for a $1 increase in marketing is $2.78.

4.1.3 HIGHER-ORDER DERIVATIVES OF MULTIVARIATE FUNCTIONS

Higher-order derivatives of multivariate functions are interpreted in the same manner as higher-order derivatives of functions of one independent variable. For $y = f(x_1, x_2)$, the second partial derivative of y with respect to x_1 describes the rate of change of the slope of the original function in planes parallel to the y and x_1 axes. Similarly, the second partial derivative of y with respect to x_2 describes the rate of change of the slope of the original function in the planes parallel to the y and x_1 axes. The symbol for the second partial derivative of y with respect to x_1 is $\delta^2 y/\delta x_1^2$, and the symbol for the second partial of y with respect to x_2 is $\delta^2 y/\delta x_2^2$.

Another higher-order derivative will be of interest in the next section. It is the *cross partial derivative*. For $y = f(x_1, x_2)$, the cross partial derivative is the derivative of $\delta y/\delta x_1$ with respect to x_2. The symbol for the cross partial derivative is $\frac{\delta^2 y}{\delta x_1\, \delta x_2}$. Alternatively, a cross partial derivative is the derivative of $\delta y/\delta x_2$ with respect to x_1. This would be written as $\frac{\delta^2 y}{\delta x_2\, \delta x_1}$. An alternative notation is f_{x_1} and $f_{x_1 x_1}$ for the first and second partial derivatives with respect to x_1 and $f_{x_1 x_2}$ or $f_{x_2 x_1}$ for the cross partial derivatives.

The cross partial is a second derivative. It describes the rate of change of the slope of the function in a given plane as the plane is shifted incrementally. Thus if $\delta y/\delta x_1$ represents the slope of the function in the family of planes that parallel the y and x_1 axes, then $\frac{\delta^2 y}{\delta x_1\, \delta x_2}$ describes the rate of change of this slope as the planes are moved marginally along the x_2 axis.

The cross partial derivatives are always equal. That is,

$$\frac{\delta^2 y}{\delta x_1\, \delta x_2} = \frac{\delta^2 y}{\delta x_2\, \delta x_1}, \quad \text{or} \quad f_{x_1 x_2} = f_{x_2 x_1}.$$

The rules for determining higher-order derivatives of functions of one independent variable apply to multivariate functions. Derivatives of multivariate functions are taken with respect to one independent variable at a time, the remaining independent variables being considered as constants.

The same procedure applies in determining higher-order derivatives of multivariate functions. The higher-order derivative is determined with respect to a single independent variable, the remaining independent variables being considered as constants. This procedure is illustrated by the following examples.

Example. Determine all first and second order derivatives of

$$y = 3x_1^3 + 2x_1^2x_2 - 3x_1x_2^2 + 4x_2^3.$$

$$\frac{\delta y}{\delta x_1} = 9x_1^2 + 4x_1x_2 - 3x_2^2$$

$$\frac{\delta y}{\delta x_2} = 2x_1^2 - 6x_1x_2 + 12x_2^2$$

$$\frac{\delta^2 y}{\delta x_1^2} = 18x_1 + 4x_2 \qquad \frac{\delta^2 y}{\delta x_2^2} = -6x_1 + 24x_2$$

$$\frac{\delta^2 y}{\delta x_1\,\delta x_2} = 4x_1 - 6x_2 \qquad \frac{\delta^2 y}{\delta x_2\,\delta x_1} = 4x_1 - 6x_2.$$

Example. Determine all first and second derivatives of

$$y = 3x_1^2 + 5x_1x_2 + 4x_2^2.$$

$$\frac{\delta y}{\delta x_1} = 6x_1 + 5x_2 \qquad \frac{\delta y}{\delta x_2} = 5x_1 + 8x_2$$

$$\frac{\delta^2 y}{\delta x_1^2} = 6 \qquad \frac{\delta^2 y}{\delta x_2^2} = 8$$

$$\frac{\delta^2 y}{\delta x_1\,\delta x_2} = 5 \qquad \frac{\delta^2 y}{\delta x_2\,\delta x_1} = 5.$$

Example. Determine all first and second derivatives of

$$y = (x_1^2 + 3x_2^3)^4.$$

$$\frac{\delta y}{\delta x_1} = 4(x_1^2 + 3x_2^3)^3(2x_1) = 8x_1(x_1^2 + 3x_2^3)^3$$

$$\frac{\delta y}{\delta x_2} = 4(x_1^2 + 3x_2^3)^3(9x_2^2) = 36x_2^2(x_1^2 + 3x_2^3)^3$$

$$\frac{\delta^2 y}{\delta x_1^2} = 8x_1[3(x_1^2 + 3x_2^3)^2(2x_1)] + 8(x_1^2 + 3x_2^3)^3$$

$$\frac{\delta^2 y}{\delta x_1^2} = 48x_1^2(x_1^2 + 3x_2^3)^2 + 8(x_1^2 + 3x_2^3)^3$$

$$\frac{\delta^2 y}{\delta x_1^2} = 8(x_1^2 + 3x_2^3)^2(7x_1^2 + 3x_2^3)$$

$$\frac{\delta^2 y}{\delta x_2^2} = 36x_2^2[3(x_1^2 + 3x_2^3)^2(9x_2^2)] + 72x_2(x_1^2 + 3x_2^3)^3$$

$$\frac{\delta^2 y}{\delta x_2^2} = 972x_2^4(x_1^2 + 2x_2^3)^2 + 72x_2(x_1^2 + 3x_2^3)^3$$

$$\frac{\delta^2 y}{\delta x_2^2} = 36x_2(x_1^2 + 3x_2^3)^2(27x_2^3 + 2x_1^2 + 6x_2^3)$$

$$\frac{\delta^2 y}{\delta x_2^2} = 36x_2(x_1^2 + 3x_2^3)^2(2x_1^2 + 33x_2^3)$$

$$\frac{\delta^2 y}{\delta x_1\, \delta x_2} = 8x_1[3(x_1^2 + 3x_2^3)^2(9x_2^2)] = 216x_1x_2^2(x_1^2 + 3x_2^3)^2$$

$$\frac{\delta^2 y}{\delta x_2\, \delta x_1} = 36x_2^2[3(x_1^2 + 3x_2^3)^2(2x_1)] = 216x_1x_2^2(x_1^2 + 3x_2^3)^2.$$

4.1.4 OPTIMUM VALUES OF MULTIVARIATE FUNCTIONS OF TWO INDEPENDENT VARIABLES

We now consider methods for determining the optimum value of functions of two independent variables. Methods of determining optimum values for functions of more than two independent variables are discussed in Sec. 4.1.5.

The optimum value of a multivariate function with two independent variables is determined in much the same manner as the optimum for a function with one independent variable. A function of two independent variables reaches a maximum or minimum when the slope of the function in both the x_1 and x_2 directions is 0. A maximum can be visualized as a hilltop and a minimum as an inverted hilltop.

A maximum or minimum can occur only where the partial derivative of the function with respect to x_1 and the partial derivative of the function with respect to x_2 are simultaneously 0. Symbolically, we state that

$$\frac{\delta f}{\delta x_1}(x_1^* = a, x_2^* = b) = 0$$

and

$$\frac{\delta f}{\delta x_2}(x_1^* = a, x_2^* = b) = 0.$$

The two equations are solved simultaneously for $x_1^* = a$ and $x_2^* = b$. The values of x_1 and x_2 for which the derivatives are 0 are again termed *critical points*.†

†They are also termed *stationary points*.

The second derivatives of the function are used to determine if a critical point is a maximum or a minimum. If both second partial derivatives are positive when evaluated at the critical points, the optimum is a minimum. If both second derivatives are negative, the optimum is a maximum.

The condition that the first derivatives equal 0 and the second derivatives are positive or negative is not sufficient to guarantee an optimum for functions of two independent variables. An additional requirement for functions of two independent variables is that the product of the second derivatives evaluated at the critical point be greater than the square of the cross partial derivative.† That is,

$$\frac{\delta^2 y}{\delta x_1^2} \cdot \frac{\delta^2 y}{\delta x_2^2} > \left[\frac{\delta^2 y}{\delta x_1 \, \delta x_2}\right]^2$$

The requirements for extrema are illustrated in the following example.

Example. For $z = 4 - x^2 - y^2$, determine the optimum. The function plots as a circular paraboloid. Note that it resembles the upper half of a football. When $z = 0$, $x^2 + y^2 = 4$, which is a circle with radius of 2. The optimum value of the function is easily seen from the sketch to be $z = 4$, and the critical point $x^* = 0$, $y^* = 0$. Using the methods discussed

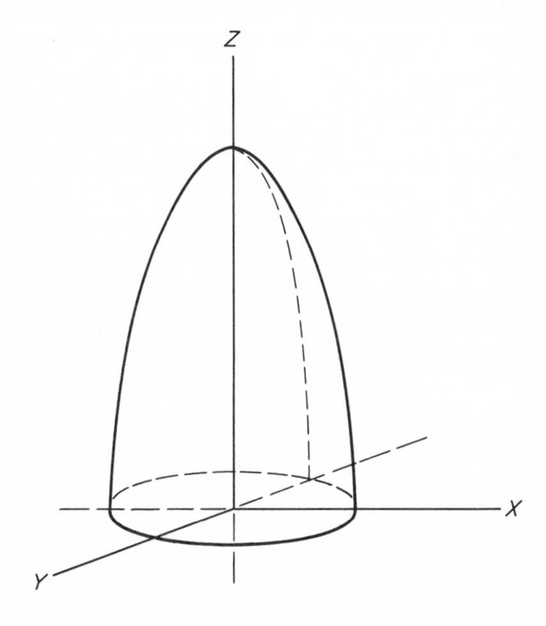

†This requirement is necessary to ensure that the function is an optimum in all directions, rather than just along the principal axis.

above, we obtain

$$\frac{\delta z}{\delta x} = -2x = 0 \quad \text{and} \quad \frac{\delta z}{\delta y} = -2y = 0.$$

Solving these equations simultaneously gives $x^* = 0$ and $y^* = 0$ as a critical point. We determine the second partial derivatives and the cross partial derivative

$$\frac{\delta^2 z}{\delta x^2} = -2, \quad \frac{\delta^2 z}{\delta y^2} = -2, \quad \text{and} \quad \frac{\delta^2 z}{\delta x\,\delta y} = 0.$$

Since

$$(-2)(-2) > (0)^2,$$

the requirement for an optimum that the product of the second derivatives be greater than the square of the cross partial derivative is satisfied. The second derivative evaluated at the critical point is negative; consequently, the optimum is a maximum. We therefore conclude that the function reaches a maximum value of $z = 4$ when $x^* = 0$ and $y^* = 0$.

In summary, the requirements for an optimum of a function $z = f(x, y)$ of two independent variables are

1. $\frac{\delta z}{\delta x}(x^* = a, y^* = b) = 0, \qquad \frac{oz}{\delta y}(x^* = a, y^* = b) = 0.$
2. $\frac{\delta^2 z}{\delta x^2}(x^* = a, y^* = b) \cdot \frac{\delta^2 z}{\delta y^2}(x^* = a, y^* = b) > \left[\frac{\delta^2 z}{\delta x\,\delta y}(x^* = a, y^* = b)\right]^2.$

Using the alternative notation for partial derivatives, we can express the requirement as

1. $f_x = 0, f_y = 0$ at (a, b).
2. $f_{xx} f_{yy} > f_{xy}^2$ at (a, b).

If these requirements are met, then for

$$\frac{\delta^2 z}{\delta x^2}(x^* = a, y^* = b) > 0,$$

the function is a minimum, and for

$$\frac{\delta^2 z}{\delta x^2}(x^* = a, y^* = b) < 0,$$

the function is a maximum.

It is possible that the product of the second derivatives evaluated at the point $x^* = a$ and $y^* = b$ will equal the square of the cross partial derivative.

When this occurs, we cannot state whether or not the function is an optimum from the above conditions. We would instead examine the points around the critical point to determine if an optimum exists.

The following examples illustrate the method for determining extreme values of functions of two independent variables.

Example. Determine the values of x and y for which $z = 4x^2 + 2y^2 + 10x - 6y - 4xy$ is an optimum, specify whether the optimum is a maximum or minimum and calculate the value of the function at the optimum.

$$\frac{\delta z}{\delta x} = 8x - 4y + 10 = 0$$

$$\frac{\delta z}{\delta y} = -4x + 4y - 6 = 0.$$

Solving simultaneously for x and y gives $x^* = -1$ and $y^* = \frac{1}{2}$.

$$\frac{\delta^2 z}{\delta x^2} = 8, \qquad \frac{\delta^2 z}{\delta y^2} = 4, \qquad \frac{\delta^2 z}{\delta x\, \delta y} = -4.$$

Since the second derivatives are positive and the product of the second derivatives is greater than the square of the cross partial derivative, the function reaches a minimum at $x^* = -1$ and $y^* = \frac{1}{2}$. The value of the function is

$$z(x^* = -1, y^* = \tfrac{1}{2}) = 4(-1)^2 - 4(-1)(\tfrac{1}{2}) + 2(\tfrac{1}{2})^2 + 10(-1) - 6(\tfrac{1}{2}) = -6\tfrac{1}{2}.$$

Example. Determine the values of x and y for which $z = -4x^2 + 4xy - 2y^2 + 16x - 12y$ is an optimum, specify whether the optimum is maximum or minimum, and calculate the value of the function at the optimum.

$$\frac{\delta z}{\delta x} = -8x + 4y + 16 = 0$$

$$\frac{\delta z}{\delta y} = 4x - 4y - 12 = 0.$$

Solving simultaneously gives $x^* = 1$ and $y^* = -2$.

$$\frac{\delta^2 z}{\delta x^2} = -8, \qquad \frac{\delta^2 z}{\delta y^2} = -4, \quad \text{and} \quad \frac{\delta^2 z}{\delta x\, \delta y} = 4.$$

Since the second derivatives are negative and the product of the second derivatives is greater than the square of the cross partial derivative, the function reaches a maximum at $x^* = 1$ and $y^* = -2$. The value of the function at the optimum is $z = 20$.

Example. Determine the critical points and specify whether the function is a maximum or minimum.

$$f(x, y) = 2x^3 - 2xy + y^2.$$

$$\frac{\delta f}{\delta x} = 6x^2 - 2y = 0$$

$$\frac{\delta f}{\delta y} = -2x + 2y = 0.$$

Substituting $2x = 2y$ into the first equation, we obtain

$$6x^2 - 2x = 0,$$

or, by factoring,

$$2x(3x - 1) = 0.$$

The expression $2x(3x - 1)$ equals 0 if $x = 0$ or if $x = \frac{1}{3}$. This means that we must investigate two critical points, namely $x^* = 0, y^* = 0$, and $x^* = \frac{1}{3}$, $y^* = \frac{1}{3}$. We determine which, if either, of these points is an optimum by the requirement

$$\frac{\delta^2 f}{\delta x^2} \cdot \frac{\delta^2 f}{\delta y^2} > \left[\frac{\delta^2 f}{\delta x\, \delta y}\right]^2.$$

For

$$\frac{\delta^2 f}{\delta x^2} = 12x, \qquad \frac{\delta^2 f}{\delta y^2} = 2, \quad \text{and} \quad \frac{\delta^2 f}{\delta x\, \delta y} = -2,$$

it is required that $(12x)(2) > [-2]^2$.
For (0, 0), $12(0)2 < (-2)^2$, the requirement is not met.
For $(\frac{1}{3}, \frac{1}{3})$, $12(\frac{1}{3})(2) > 4$, the requirement is met.
For $(\frac{1}{3}, \frac{1}{3})$, determine whether $f(x, y)$ is maximum or minimum.

$$\frac{\delta^2 f}{\delta x^2} = 12x \quad \text{and} \quad 12(\tfrac{1}{3}) > 0,$$

the optimum is a minimum.

It is possible for a function of two independent variables to reach a maximum in the plane of one pair of axes and a minimum in another. Fig. 4.2 shows a graph of the function $z = x^2 - y^2$. z reaches a maximum at the origin in the plane formed by the z and y axes but simultaneously reaches a minimum in the plane formed by the z and x axes. The point $(x^* = 0, y^* = 0)$ for this function is neither a maximum nor a minimum; it is called a *saddle point*.

We have stated that an optimum occurs at the critical points, provided that the product of the second derivatives exceeds the square of the cross partial derivative. If the square of the cross partial derivative exceeds the product of the second derivatives, the function has reached a saddle point.

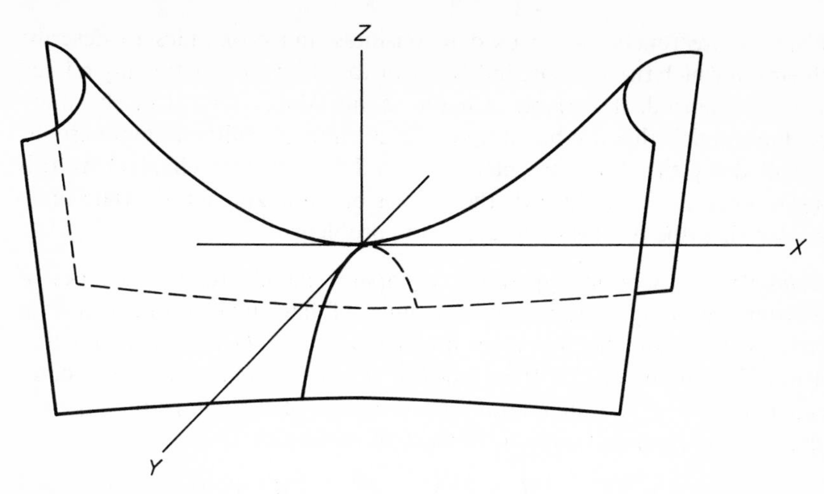

Figure 4.2

A function that reaches a saddle point is illustrated by the following example.

Example. Verify that a saddle point exists for the function

$$z = 18x^2 - 6y^2 - 36x - 48y.$$

We first determine the critical values of x and y by equating the partial derivatives of z with respect to x and y with 0 and solving the pair of equations simultaneously for x^* and y^*.

$$\frac{\delta z}{\delta x} = 36x - 36 = 0$$

$$\frac{\delta z}{\delta y} = -12y - 48 = 0.$$

Solving for the critical values of x and y gives $x^* = 1$ and $y^* = -4$. The second derivatives of z are

$$\frac{\delta^2 z}{\delta x^2} = 36, \quad \frac{\delta^2 z}{\delta y^2} = -12, \quad \text{and} \quad \frac{\delta^2 z}{\delta x\, \delta y} = 0.$$

The second derivatives are of different sign. The requirement for a saddle point that

$$\frac{\delta^2 z}{\delta x^2} \cdot \frac{\delta^2 z}{\delta^2 y} < \left[\frac{\delta^2 z}{\delta x\, \delta y}\right]^2$$

is satisfied. We therefore conclude that the function reaches a saddle point at $x^* = 1$ and $y^* = -4$.

Multivariate functions are used in business and economics to describe problems in which two or more independent variables are functionally related to a single dependent variable. Chapter 5 introduces several of the more important models that are based upon the calculus of multivariate functions. We shall delay the development of the models until that chapter. At this point, however, we can illustrate the process of optimizing multivariate functions that describe business and economic problems.

Example. The Corola Typewriter Company manufactures two types of typewriters. Both typewriters are machines of fine quality, one being an electric portable and the second a manual portable. The revenue function that describes total revenue from sales of the two machines is rather complicated because of the interaction between sales of the competing types of machines. The revenue function, in units of thousands, is

$$R = 8E + 5M + 2EM - E^2 - 2M^2 + 20,$$

where E represents the units of electric in thousands and M represents units of manual in thousands. Determine the quantity of electric and manual typewriters which leads to maximum revenue.

To determine the critical values of E and M we equate the partial derivatives of the revenue function with 0.

$$\frac{\delta R}{\delta E} = 8 + 2M - 2E = 0$$

$$\frac{\delta R}{\delta M} = 5 - 4M + 2E = 0.$$

Solving simultaneously for E and M gives

$$E^* = 10.5 \quad \text{and} \quad M^* = 6.5.$$

Revenue is maximum when 10,500 electric and 6500 manual typewriters are manufactured. The maximum revenue is $R =$ \$78,250.

The second derivatives are negative and their product exceeds the square of the cross partial derivatives. This indicates that the critical values of the variables represent a maximum rather than a minimum or an inflection point.

The Corola Typewriter Company offers an example in which two products have a degree of substitutability between each other. The example that follows is based upon the same principle, with the exception that inputs to the production process are considered rather than the output from the production process. The objective is to maximize output by using the appropriate quantities of factors of production.

Example. Western Valley Farms, Inc. has developed a sales revenue function that treats farm equipment E and labor L as independent variables with sales revenue as the dependent variable. The revenue function is

$$R = 18L + 24E + 10LE - 5E^2 - 8L^2,$$

where L represents units of labor and E represents units of farm equipment. Determine the amount of labor and equipment that gives maximum revenue.

The partial derivatives are

$$\frac{\delta R}{\delta L} = 18 + 10E - 16L = 0$$

$$\frac{\delta R}{\delta E} = 24 - 10E + 10L = 0.$$

Solving the two equations simultaneously gives

$$L^* = 7 \quad \text{and} \quad E^* = 9.4.$$

The maximum revenue from the inputs to the production process is $R = \$176$. Since the second derivatives are negative and since the product of the second derivatives exceeds the square of the cross partial derivative, the revenue is maximum for these inputs.

4.1.5 OPTIMUM VALUES OF MULTIVARIATE FUNCTIONS OF MORE THAN TWO INDEPENDENT VARIABLES†

Optimum values of functions of more than two variables are determined by equating the partial derivatives to zero and solving the resulting equations simultaneously for the critical values of each independent variable. Thus, for $y = f(x_1, x_2, \ldots, x_n)$, the procedure for determining the critical points is to equate the partial derivatives to zero and to solve the n partial derivative equations simultaneously for the critical points. That is,

$$\frac{\delta f}{\delta x_1}(x_1^*, x_2^*, \ldots, x_n^*) = 0$$

$$\frac{\delta f}{\delta x_2}(x_1^*, x_2^*, \ldots, x_n^*) = 0$$

$$\frac{\delta f}{\delta x_n}(x_1^*, x_2^*, \ldots, x_n^*) = 0.$$

The rules for specifying if a critical point is a maximum, a minimum, or some combination of maximum and minimum are beyond the scope of this text.‡ We can, however, determine if the function is a maximum or minimum in specific planes.

---

†This section can be omitted without loss of continuity.

‡The technique is discussed by Alpha C. Chiang, *Fundamental Methods of Mathematical Economics* (McGraw-Hill, 1967), pp. 232–235.

Consider the function $y = f(x_1, x_2, x_3)$. The critical points are determined by solving the following equations simultaneously.

$$\frac{\delta f}{\delta x_1} = 0, \quad \frac{\delta f}{\delta x_2} = 0, \quad \frac{\delta f}{\delta x_3} = 0.$$

For the critical point (x_1^*, x_2^*, x_3^*), we can state that the function is a maximum in the planes parallel to the y and x_1 axes, the y and x_2 axes, and the y and x_3 axes provided that the second derivatives are negative. The function is a minimum in these planes if the second derivatives are positive.

To summarize, for a multivariate function to achieve an optimum it is necessary that all first derivatives be zero and all second derivatives (excluding cross partials) have the same sign. These conditions, however, are not sufficient. A third requirement, which is beyond the scope of this text, must be met. The student who takes advanced studies will be introduced to the requirement. The student who is not interested in advanced mathematics need only remember that functions of three or more independent variables cannot be treated with the same ease as those with only two independent variables.

The method of determining optimum values of a multivariate function along the major axes is illustrated in the following example.

Example. Determine the critical points for the following functions. Specify whether the function reaches a maximum or minimum along the major axes at the critical point.

$$f(x_1, x_2, x_3) = 2x_1^2 + x_1x_2 + 4x_2^2 + x_1x_3 + x_3^2 + 2.$$

$$\frac{\delta f}{\delta x_1} = 4x_1 + x_2 + x_3 = 0$$

$$\frac{\delta f}{\delta x_2} = x_1 + 8x_2 = 0$$

$$\frac{\delta f}{\delta x_3} = x_1 + 2x_3 = 0.$$

Solving simultaneously gives $x_1^* = 0$, $x_2^* = 0$, and $x_3^* = 0$ as the critical point. The second derivatives are

$$\frac{\delta^2 f}{\delta x_1^2} = 4, \quad \frac{\delta^2 f}{\delta x_2^2} = 8, \quad \frac{\delta^2 f}{\delta x_3^3} = 2.$$

Since each of the second derivatives are positive, we conclude that the function reaches a minimum in the major axes at the critical point $x_1^* = 0$, $x_2^* = 0$, and $x_3^* = 0$.

4.2 exponential and logarithmic functions

Our discussion in Chapter 1 was centered on algebraic functions. Another important class of functions that are widely used in business and economic models are the exponential and logarithmic functions. The exponential and logarithmic functions are subclassifications of a more general classification of functions termed *transcendental functions.* Since the exponential and logarithmic are the most widely applicable of the transcendental functions to business and economic problems, we shall limit our discussion to these two functional forms.

4.2.1 EXPONENTIAL FUNCTION

An exponential function is a function of the form

$$f(x) = k(a)^{cx}, \tag{4.4}$$

where $f(x)$ is the dependent variable, x is the independent variable, a is a constant termed the *base* that is greater than 0 and not equal to 1, and k and c are parameters that position the function on the coordinate axes. The possible forms of the exponential function are graphed in Fig. 4.3.

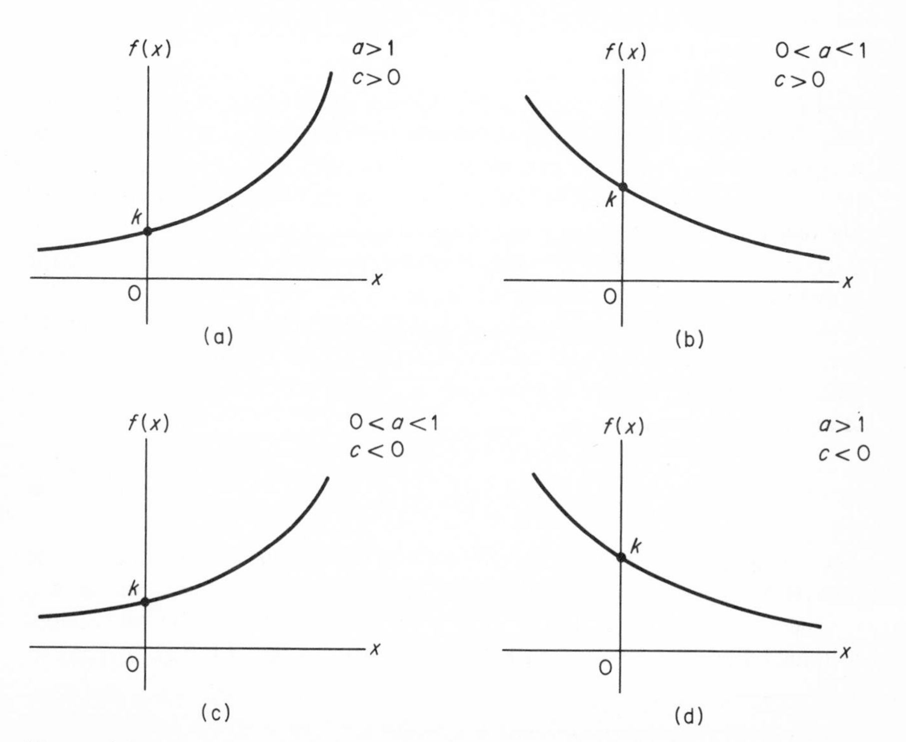

Figure 4.3

Figures 4.3(a) and 4.3(c) both show an exponential function in which the dependent variable increases as the independent variable increases. It can be shown that by proper selection of the constants a and c in Fig. 4.3(a), the functional relationship in Fig. 4.3(c) is completely described. Similarly, the functional relationships shown in Figs. 4.3(b) and 4.3(d) are interchangeable. The most commonly used forms of the exponential function are those in which a is greater than 1, the functions shown in Figs. 4.3(a) and 4.3(d). Those shown in Figs. 4.3(b) and 4.3(c) are occasionally applied in business and economic problems.

One of the characteristics of the exponential function that finds widespread application in business and economic problems is that the function describes constant rates of growth. As the independent variable increases by a constant amount in the exponential function, the dependent variable increases or decreases by a constant percentage. Thus, the value of an investment that increases by a constant percentage each period, the sales of a company that increase at a constant rate each period, and the value of an asset that declines at a constant rate each period are examples of functional relationships that are described by the exponential function.

As an example of an exponential function, consider the function

$$A = P(1 + i)^n. \tag{4.5}$$

This function is commonly used to describe the amount of money A that will accrue if an original principal P is invested for n periods and earns interest or grows at the rate i per period. If, for example, it is anticipated that an investment of \$1000 in securities will grow at an annual rate of 10 percent per year for ten years, the amount of the investment at the end of the nth year is given by $A = 1000(1.10)^n$. This function, graphed in Fig. 4.4, shows the growth of P given the parameters i and n.

The value of A for each period may be calculated by multiplying the accrued amount at the end of the previous period by the factor $(1 + i)$. The iterative formula for determining A_n, where A_n represents the value of A at the end of period n, is

$$A_n = A_{n-1}(1 + i). \tag{4.6}$$

The calculations for the amount for each of the first ten years if we assume interest of 10 percent and an original principal of \$1000 are shown in Fig. 4.4.

The value of A can also be calculated from (4.5) with the use of logarithms.† For $n = 20$, $P =$ \$1000, and $i =$ 10 percent, A is \$6730. The cal-

†The use of logarithms is explained in Appendix B.

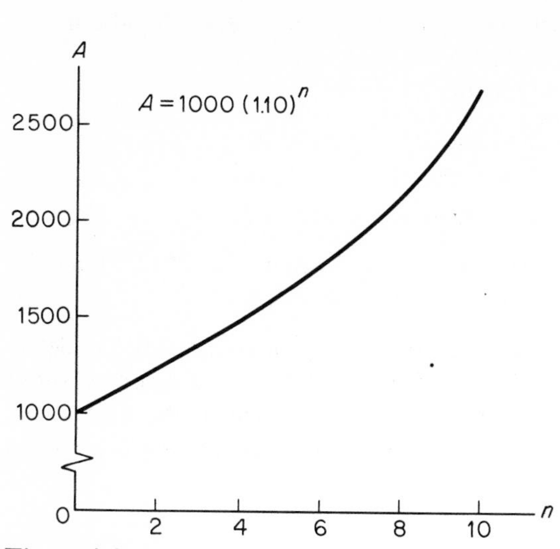

Figure 4.4

n	A
0	1000.00
1	1100.00
2	1210.00
3	1331.00
4	1464.10
5	1610.51
6	1771.56
7	1948.72
8	2143.59
9	2357.95
10	2593.75

culations are

$$\begin{aligned} A &= \$1000\,(1.10)^{20} \\ \log A &= \log(1000) + 20 \log(1.10) \\ \log A &= 3.0000 + 20(0.0414) \\ \log A &= 3.8280 \\ A &= \$6730. \end{aligned}$$

The following examples illustrate the use of exponential functions in describing constant rates of growth.

Example. The cost of sending a student to a certain school has been increasing at an average rate of 6 percent per year. If the present cost is \$1500, determine the cost in 10 years.

The functional relationship is

$$\begin{aligned} A &= 1500\,(1.06)^{10} \\ \log A &= \log(1500) + 10 \log(1.06) \\ \log A &= 3.1761 + 10(0.0253) \\ \log A &= 3.4291. \end{aligned}$$

Thus,

$$A = \$2686.$$

Example. Profits during the past three years have increased at a rate of 12 percent. If profits during the past year were \$750,000, develop a formula for predicting profits that assumes a continuation of the growth pattern, and use this formula to project profits for the next three years.

Applying formula (4.6), we obtain

$$\begin{aligned} \text{Profits}_1 &= \$750{,}000(1.12) = \$840{,}000 \\ \text{Profits}_2 &= \$840{,}000(1.12) = \$940{,}000 \\ \text{Profits}_3 &= \$940{,}000(1.12) = \$1{,}054{,}560. \end{aligned}$$

This formula yields profits of \$840,000 for the current year, \$940,800 for the following year, and \$1,054,560 for the third year.

Example. A company that manufactures copying equipment has determined that the life of a copier is limited by obsolescence. If a copier whose initial value is \$10,000 decreases in value by 20 percent of its value at the beginning of the preceding year, determine the value of the machine at the end of each of the five years.

The functional relationship is

$$A = \$10{,}000(0.80)^n.$$

The value at the end of each year is given in the following table. The calculations were made by using formula (4.5) and logarithms.

n	1	2	3	4	5
A	8000	6400	5120	4096	3277

4.2.2 THE NUMBER *e*

One of the most commonly used bases in the exponential function is the number e. e is defined as the limit as n approaches infinity of $(1 + 1/n)^n$. That is,

$$e = \lim_{n \to \infty} \left(1 + \frac{1}{n}\right)^n. \tag{4.7}$$

The numerical value of e with accuracy of nine places is 2.718281828. Replacing a by e in the formula for an exponential function, we obtain

$$f(x) = ke^{cx}, \tag{4.8}$$

where k and c are again parameters, $f(x)$ and x are the dependent and independent variables, and e is the base of the function and has the value 2.718281828.

The popularity and widespread usage of e as the base in the exponential function stems partially from properties of the derivative of this function.

Before introducing rules for determining derivatives of exponential functions, however, it will be useful to introduce methods of expressing the growth function in terms of base e rather than base a.

The function $A = P(1 + i)^n$ can be expressed with the base e by the following transformation. Let

$$P(1 + i)^n = Pe^{cn},$$

or

$$(1 + i)^n = e^{cn}.$$

By taking the natural logarithm of the expressions on both sides of the equal sign, we obtain

$$n \ln (1 + i) = cn$$

and

$$c = \ln (1 + i).$$

Thus, the growth function expressed in terms of the base e is:

$$A = Pe^{\ln(1+i)n}. \tag{4.9}$$

This function can be easily evaluated through use of Table IV of natural logarithms. The procedure is illustrated in the following examples.

Example. The market value of real estate in a certain area has increased by 18 percent per year. If the present value of the real estate is \$15,000, what is the anticipated value four years hence?

To simplify the calculations, we first express \$15,000 as $\$15 \cdot 10^3$. This enables us to determine the natural logarithm of 15 by the use of Table IV. We retain 1000 as 10^3 for the calculations.

If

$$\begin{aligned} A &= \$15e^{\ln(1.18)4}(10^3) \\ \ln A &= \ln (15) + 4[\ln (1.18)] + \ln (10^3) \\ \ln A &= 2.70805 + 4(0.16551) + \ln (10^3) \\ \ln A &= 3.37009 + \ln (10^3), \end{aligned}$$

then

$$A = 29.1(10^3) = \$29,100.$$

Example. In a moment of weakness, John Smith promised his wife Mary that they would take an "around the world" vacation trip beginning on their tenth wedding anniversary. John anticipates the trip's costing \$5000. John and Mary were married last month. What amount would John need to deposit at interest of 5 percent in order to pay for the trip?

If

$$5000 = Pe^{\ln(1.05)10},$$

then

$$P = \frac{5000}{e^{\ln(1.05)10}} = \frac{5 \cdot 10^3}{e^{\ln(1.05)10}}$$

and

$$\begin{aligned} \ln P &= \ln 5 - 10 \ln (1.05) + \ln (10^3) \\ \ln P &= 1.60944 - 10(0.04879) + \ln (10^3) \\ \ln P &= 1.12154 + \ln (10^3) \\ P &= 3.07(10^3) = \$3070. \end{aligned}$$

Example. Company A has stated that it expects sales to increase from the present level of $1,000,000 by $50,000 per year. Company B expects an annual growth rate of 6 percent and it reported sales of $900,000 during the past accounting period. Develop predicting formulas for both companies and calculate anticipated sales 10 years hence.

For Company A:

$$S = 1{,}000{,}000 + 50{,}000t$$

and

$$S(10) = 1{,}000{,}000 + 500{,}000 = \$1{,}500{,}000.$$

For Company B:

$$S = \$900{,}000e^{\ln(1.06)t}$$

and

$$\begin{aligned} S(10) &= \$900{,}000e^{10 \ln(1.06)} \\ &= \$1{,}610{,}000. \end{aligned}$$

4.2.3 DERIVATIVES OF EXPONENTIAL FUNCTIONS

The derivative of an exponential function gives the slope or rate of change of the function for values of the independent variables. There are two rules for determining the derivatives of exponential functions. One applies to exponential functions in base e, the other to exponential functions with base other than e.

Rule 11. If $y = e^{g(x)}$, where $g(x)$ is a differentiable function of x, then

$$\frac{dy}{dx} = e^{g(x)}g'(x).$$

Rule 12. If $y = a^{g(x)}$, where $g(x)$ is a differentiable function of x, then

$$\frac{dy}{dx} = a^{g(x)}g'(x) \ln a.$$

These two rules are illustrated by the following examples.

Example. $y = e^x$; determine dy/dx.

$$\frac{dy}{dx} = e^x(1).$$

The derivative of e^x is also e^x.

Example. $y = 2^x$; determine dy/dx.

$$\frac{dy}{dx} = 2^x(1)\ln 2 = 2^x(0.69315).$$

Example. $y = 3e^{5x}$; determine the slope of the function at $x = 0.5$.

$$\frac{dy}{dx} = 3e^{5x}(5) = 15e^{5x}$$

$$\frac{dy}{dx}(x = 0.5) = 15e^{2.5} = 15(12.18) = 182.7.$$

The slope of the function at $x = 0.5$ is 182.7. The value of 12.18 for $e^{2.5}$ was obtained from Table V.

Example. $y = 4(3)^{-0.1x}$; determine dy/dx.

$$\frac{dy}{dx} = 4(3)^{-0.1x}(-0.1)\ln 3.$$

Example. $y = xe^{x^2}$; determine dy/dx.

$$\frac{dy}{dx} = x[e^{x^2}(2x)] + e^{x^2}(1).$$

This result comes from the product rule and the exponential rule.

Example. $y = xe^{2x}$; determine the optimum value of the function and specify whether the function is a maximum or a minimum.

$$\frac{dy}{dx} = xe^{2x}(2) + e^{2x}(1) = 0$$

$$e^{2x}(2x + 1) = 0.$$

In order for the expression to equal 0, either e^{2x} equals 0 or $(2x + 1)$ equals 0. There is no value for x such that e^{2x} equals 0. Therefore, the critical value of x is determined by equating $2x + 1$ with 0. The function is an optimum when $x^* = -\frac{1}{2}$. To determine if the critical point is a maximum or minimum, we evaluate the second derivative at the critical point.

$$\frac{d^2y}{dx^2} = e^{2x}(2) + (2x + 1)e^{2x}(2)$$

$$= 2e^{2x}(2x + 2)$$

$$\frac{d^2y}{dx^2}\left(x^* = -\frac{1}{2}\right) = 2e^{-1}(-1 + 2) = \frac{2}{e} > 0.$$

Since the second derivative is positive, the function is a minimum at $x^* = -\frac{1}{2}$.

Example. $y = e^{(2x_1+x_2)}$; determine the first derivatives.

$$\frac{\delta y}{\delta x_1} = e^{(2x_1+x_2)}(2) = 2e^{(2x_1+x_2)}$$

$$\frac{\delta y}{\delta x_2} = e^{(2x_1+x_2)}(1) = e^{(2x_1+x_2)}.$$

Example. $y = x_1x_2e^{(x_1+2x_2)}$; determine the first derivatives.

$$\frac{\delta y}{\delta x_1} = (x_1x_2)e^{(x_1+2x_2)}(1) + e^{(x_1+2x_2)}[x_2]$$

$$\frac{dy}{dx_1} = e^{x_1+2x_2}(x_1x_2 + x_2)$$

$$\frac{\delta y}{\delta x_2} = x_1x_2e^{(x_1+2x_2)}(2) + e^{x_1+2x_2}[x_1]$$

$$\frac{dy}{dx_2} = e^{(x_1+2x_2)}(2x_1x_2 + x_1).$$

Example. The demand function for a commodity is described by the exponential function $p = 7.50e^{-0.001q}$. Determine the quantity and price for which total revenue is a maximum.

$$TR = q \cdot p = q(7.50e^{-0.001q})$$

$$\frac{d(TR)}{dq} = 7.50q[e^{-0.001q}(-0.001)] + 7.50e^{-0.001q} = 0$$

$$= 7.50e^{-0.001q}(-0.001q + 1) = 0.$$

We must determine the value of q for which the expression is 0. For the expression to equal 0, either $7.50e^{-0.001q}$ equals 0 or $-0.001q + 1$ equals 0. Since $7.50e^{-0.001q}$ is not equal to 0 for any value of q, we equate $-0.001q + 1$ with 0 and obtain $q = 1000$.

4.2.4 LOGARITHMIC FUNCTIONS

The logarithmic function is the inverse of the exponential function. In Sec. 1.2.4, we explained that if $y = f(x)$ was solved for x to give $x = g(y)$, then $y = f(x)$ and $x = g(y)$ are inverse functions.

It can be shown that the logarithmic and exponential functions are inverse functions. Given the exponential function $y = e^x$, one can obtain $x = \ln y$ by taking the logarithm of both sides of the exponential function. Thus, $y = e^x$ and $x = \ln y$ are inverse functions.

The logarithmic function describes a functional relationship in which constant proportional changes in the independent variable result in constant arithmetic changes in the dependent variable. As an example, consider the exponential function $y = 10^x$. As x increases in units of 1, y increases by a factor of 10. If this exponential function is rewritten as a logarithmic function, we obtain $x = \log y$. In the logarithmic form, for $y = 10$, $x = 1$, for $y = 100$, $x = 2$, etc. We conclude that constant proportional changes in the independent variable y result in constant arithmetic changes in x. Fig. 4.5 provides a graph of the logarithmic function.

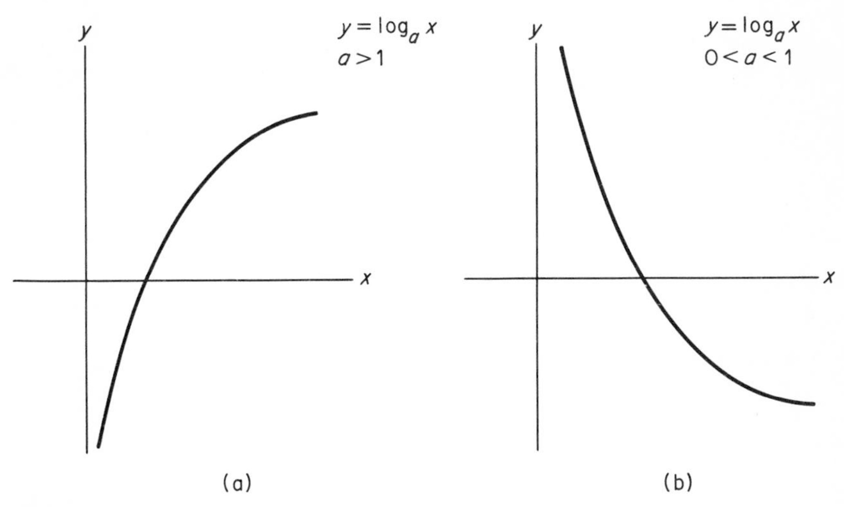

Figure 4.5

There are two rules for determining the derivatives of logarithmic functions. Rule 13 applies to logarithmic functions of the base e, and Rule 14 applies to all bases other than e.

Rule 13. If $y = \ln g(x)$, where $g(x)$ is a differentiable function of x, then

$$\frac{dy}{dx} = \frac{g'(x)}{g(x)}.$$

Rule 14. If $y = \log_a g(x)$, where $g(x)$ is a differentiable function of x, then

$$\frac{dy}{dx} = \frac{g'(x)}{g(x)} \cdot \log_a e.$$

Example. $y = \ln x$; determine dy/dx.

$$\frac{dy}{dx} = \frac{1}{x}.$$

Example. $y = \log_{10} x$; determine dy/dx.

$$\frac{dy}{dx} = \frac{1}{x} \log_{10} e = \frac{0.4343}{x}.$$

Example. $y = \ln(x^2 + 2x + 10)$; determine dy/dx.

$$\frac{dy}{dx} = \frac{2x + 2}{x^2 + 2x + 10}.$$

Example. $y = [\ln(x + 3)]^2$; determine dy/dx.

$$\frac{dy}{dx} = 2[\ln(x + 3)]^1 \frac{1}{x + 3}.$$

Example. Determine the slope of the function in the preceding example at $x = 2$.

$$\frac{dy}{dx}(x = 2) = \frac{2 \ln 5}{5} = \frac{2}{5}(1.60944) = 0.6438.$$

Example. $y = \ln(x^2 - 4x + 6)$; determine the critical point. Specify whether the function is a maximum or minimum at the critical point.

$$\frac{dy}{dx} = \frac{2x - 4}{x^2 - 4x + 6} = 0.$$

Multiplying both sides by $x^2 - 4x + 6$ gives

$$2x - 4 = 0,$$

and $x^* = 2$ is a critical point. Evaluating the second derivative at $x^* = 2$ gives

$$\frac{d^2y}{dx^2} = \frac{(x^2 - 4x + 6)(2) - (2x - 4)(2x - 4)}{(x^2 - 4x + 6)^2},$$

$$\frac{d^2y}{dx^2}(x = 2) = 1 > 0.$$

Thus, the function reaches a minimum at $x^* = 2$.

Example. $y = \ln(x_1^2 + 2x_1x_2)$; determine the partial derivatives.

$$\frac{\delta y}{\delta x_1} = \frac{2x_1 + 2x_2}{x_1^2 + 2x_1x_2}$$

$$\frac{\delta y}{\delta x_2} = \frac{2x_1}{x_1^2 + 2x_1x_2}.$$

4.2.5 LOGARITHMIC DIFFERENTIATION

The derivatives of certain functions can be most easily determined if the function is expressed in logarithms. This technique, termed *logarithmic differentiation,* is especially applicable to functions of the form

$$y = f(x)^{g(x)}. \tag{4.10}$$

As an example, consider the function $y = x^x$. Neither the rule for the derivative of a variable raised to a constant, Rule 3, nor that for a constant raised to a variable, Rule 12, applies to this function. Consequently, we must develop a formula that gives the derivative of a variable raised to a variable power.

In developing the rule, we first take the natural logarithm of both sides of (4.10).

$$\ln y = g(x)[\ln f(x)].$$

To illustrate the derivative process for this function, we shall consider the derivative of the term on the left side of the equal sign and that on the right side separately. The derivative of $\ln y$ with respect to x is found by using Rule 13:

$$\frac{d(\ln y)}{dx} = \frac{1}{y}\frac{dy}{dx}\cdot$$

By the application of the product rule, Rule 6, and Rule 13, the derivative of $g(x)[\ln f(x)]$ is

$$\frac{d[g(x)\ln f(x)]}{dx} = g(x)\frac{f'(x)}{f(x)} + g'(x)\ln f(x).$$

The derivative of

$$\ln y = g(x)\ln f(x)$$

is thus

$$\frac{1}{y}\frac{dy}{dx} = g(x)\frac{f'(x)}{f(x)} + g'(x)\ln f(x).$$

Multiplying both sides by y gives

$$\frac{dy}{dx} = y\left[g(x)\frac{f'(x)}{f(x)} + g'(x)\ln f(x)\right]\cdot$$

If we now replace y by $f(x)^{g(x)}$, we obtain a formula for determining the derivative of a variable raised to a variable power.

Rule 15. If $y = f(x)^{g(x)}$, the derivative is given by

$$\frac{dy}{dx} = f(x)^{g(x)}\left[g(x)\frac{f'(x)}{f(x)} + g'(x)\ln f(x)\right]\cdot$$

Example. $y = x^x$; determine dy/dx.

$$\frac{dy}{dx} = x^x\left[\frac{x(1)}{x} + (1)\ln x\right] = x^x(1 + \ln x).$$

Example. $y = (x + 4)^{x^2}$; determine dy/dx.

$$\frac{dy}{dx} = (x + 4)^{x^2}\left[x^2\frac{1}{(x + 4)} + 2x\ln(x + 4)\right]\cdot$$

The technique used in the derivation of Rule 15 often proves useful in determining derivatives of complicated functions other than those of the form shown by (4.10). This technique is illustrated by the following examples.

Example. $y = \dfrac{(x-2)^3}{(x+3)^4}$; determine dy/dx.

First express the function in terms of natural logarithms.

$$\ln y = 3\ln(x-2) - 4\ln(x+3).$$

The derivative of this function is

$$\frac{1}{y}\frac{dy}{dx} = \frac{3}{x-2} - \frac{4}{x+3} = \frac{-(x-17)}{(x+3)(x-2)}$$

$$\frac{dy}{dx} = -\frac{(x-2)^3}{(x+3)^4}\left[\frac{(x-17)}{(x+3)(x-2)}\right] = -\frac{(x-2)^2(x-17)}{(x+3)^5}.$$

This result can be verified by applying the quotient rule, Rule 7, to the original function.

Example. $y = \dfrac{x^3(1+x^2)^{1/2}}{(x^2-5)^2}$; determine dy/dx.

$$\ln y = 3\ln x + \tfrac{1}{2}\ln(1+x^2) - 2\ln(x^2-5).$$

$$\frac{1}{y}\frac{dy}{dx} = \frac{3}{x} + \frac{1}{2}\frac{2x}{(1+x^2)} - \frac{2(2x)}{(x^2-5)}$$

$$\frac{dy}{dx} = \frac{x^3(1+x^2)^{1/2}}{(x^2-5)^2}\left[\frac{3}{x} + \frac{x}{(1+x^2)} - \frac{4x}{(x^2-5)}\right].$$

4.3 constrained optima

We shall often be faced with the problem of maximizing profit subject to a constraint such as the availability of working capital. Alternatively, a common problem requires the maximization of sales revenue subject to a constraint on production capacity. The problem of determining the optimum value of a *quadratic* or *higher-order function* subject to one or more *linear* or *higher-order constraints* can be solved by using differential calculus.

The general problem is that of finding the extreme points of the multivariate function $z = f(x, y)$ subject to equalities of the form $g(x, y) = 0$. The function $z = f(x, y)$ is termed the *objective function* and $g(x, y) = 0$ is the *constraining equation*. If we represent the critical values of x and y by x^* and y^*, the problem becomes that of determining critical points such that

$$\frac{\delta f}{\delta x}(x^*, y^*) = 0, \qquad \frac{\delta f}{\delta y}(x^*, y^*) = 0, \quad \text{and} \quad g(x^*, y^*) = 0.$$

4.3.1 SUBSTITUTION

One method of solving the constrained optima problem is that of substitution. This involves solving the constraining equation for one variable in terms of the other and substituting this expression in the objective function. The objective function is then optimized by using the procedure discussed in the preceding section. This is illustrated by the following example.

Example. Determine the local optima of $z = x^2 + 3xy + y^2$, subject to the constraining equation $x + y = 100$.

Solving the constraining equation for y and substituting into the objective function, we obtain

$$z = x^2 + 3x(100 - x) + (100 - x)^2.$$

To determine the critical point we equate dz/dx with 0 and solve for x.

$$\frac{dz}{dx} = 2x + 3x(-1) + 3(100 - x) + 2(100 - x)(-1) = 0,$$

and

$$x^* = 50, \qquad y^* = 100 - x^* = 50$$

are the critical points.

Applying the second derivative test to the function gives

$$\frac{d^2z}{dx^2} = -2.$$

This indicates that the constrained function is a maximum in the plane parallel to the z and x axes. If the objective function is written in terms of y rather than x, we find that the function also reaches a local maximum in the plane parallel to the z and y axes. The value of the constrained maximum is $z = 12{,}500$.

The method of substitution in solving constrained optima problems becomes quite difficult if the constraint is at all complex. For example, the method of substitution is rather cumbersome because of the difficulty of solving for x in terms of y for constraints of the form $ax^2 + bxy + y^2 = 0$. An alternative method of solution is, however, available. This is the method of Lagrangian multipliers.

4.3.2 LAGRANGIAN MULTIPLIERS

Lagrangian multipliers provide a method of optimizing quadratic or higher-order functions subject to linear or higher-order constraints. The method of solution is to form the Lagrangian expression,

$$F(x, y, \lambda) = f(x, y) + \lambda g(x, y). \qquad (4.11)$$

This expression consists of the objective function $f(x, y)$, and the product of λ (lambda) and the constraining equation $g(x, y)$. The coefficient of the constraining equation, λ, is termed the Lagrangian multiplier. Since the constraining equation is equal to 0, the addition of the term $\lambda g(x, y)$ to the objective function $f(x, y)$ does not change the value of the function. The necessary condition for an extreme point is that the partial derivatives of the function with respect to x, y, and λ equal 0; that is,

$$\frac{\delta F}{\delta x} = 0, \quad \frac{\delta F}{\delta y} = 0, \quad \text{and} \quad \frac{\delta F}{\delta \lambda} = 0.$$

The three equations with variables x, y, and λ are solved simultaneously for the critical values of x^*, y^*, and λ^*.

Since the Lagrangian expression has more than two independent variables, the second derivative test does not apply in specifying whether the function has reached a local maximum or minimum. Rather, we can determine if the function is a local maximum or minimum by evaluating the function at points adjacent to the critical points in the constraining plane. The procedure is summarized as follows. Let x^* and y^* represent critical points in the Lagrangian expression. $f(x^*, y^*)$ is a constrained maximum if

$$f(x^*, y^*) > f(x^* - \Delta x, y^* + \Delta y),$$

and

$$f(x^*, y^*) > f(x^* + \Delta x, y^* - \Delta y).$$

It is a constrained minimum if

$$f(x^*, y^*) < f(x^* - \Delta x, y^* + \Delta y),$$

and

$$f(x^*, y^*) < f(x^* + \Delta x, y^* - \Delta y),$$

where both $g(x^* + \Delta x, y^* - \Delta y) = 0$, and $g(x^* - \Delta x, y^* + \Delta y) = 0$. In selecting Δx and Δy, Δx need not equal Δy. However, it is important to remember that Δx and Δy must be selected such that the constraining equation remains equal to 0.

The method of Lagrangian multipliers is illustrated by the following examples.

Example. Determine the critical points and the constrained optima for

$$z = x^2 + 3xy + y^2$$

subject to

$$x + y = 100.$$

The Lagrangian expression is

$$F(x, y, \lambda) = x^2 + 3xy + y^2 + \lambda(x + y - 100),$$

and the partial derivatives are

$$\frac{\delta z}{\delta x} = 2x + 3y + \lambda = 0,$$

$$\frac{\delta z}{\delta y} = 3x + 2y + \lambda = 0,$$

$$\frac{\delta z}{\delta \lambda} = x + y - 100 = 0.$$

The three equations are solved simultaneously for

$$x^* = 50, \qquad y^* = 50, \quad \text{and} \quad \lambda^* = -250.$$

The critical values of x, y, and λ can be substituted into the Lagrangian expression to obtain the constrained optimum. For the example problem,

$$\begin{aligned} f(50, 50, -250) &= (50)^2 + 3(50)(50) + (50)^2 - 250(50 + 50 - 100) \\ &= 2500 + 7500 + 2500 - 250(0) \\ &= 12{,}500. \end{aligned}$$

The function is a constrained maximum, since

$$F(49, 51, -250) = 12{,}499,$$

and

$$F(51, 49, -250) = 12{,}499.$$

Example. Determine the critical points and the constrained optima for

$$z = x^2 - xy + y^2$$

subject to

$$x + y = 100.$$

The Lagrangian expression is

$$F(x, y, \lambda) = x^2 - xy + y^2 + \lambda(x + y - 100),$$

and the partial derivatives are

$$\frac{\delta F}{\delta x} = 2x - y + \lambda = 0$$

$$\frac{\delta F}{\delta y} = -x + 2y + \lambda = 0$$

$$\frac{\delta F}{\delta \lambda} = x + y - 100 = 0.$$

The three equations are solved simultaneously for

$$x^* = 50, \qquad y^* = 50, \quad \text{and} \quad \lambda = -50.$$

The constrained optimum is

$$F(50, 50, -50) = (50)^2 - (50)(50) + (50)^2 - 50(0)$$
$$= 2500.$$

This constrained optimum is a local minimum, since

$$F(49, 51, -50) = 2503,$$

and

$$F(51, 49, -50) = 2503.$$

In this and the preceding example the constraining equation $g(x, y) = x + y - 100 = 0$ is satisfied by the data points (49, 51) and (51, 49). We have, therefore, been able to use these points to determine if the optimum was a constrained maximum or constrained minimum.

As stated previously, the second derivative test does not apply in Lagrangian multipliers. This is illustrated by the preceding two examples. In both examples the second derivatives were positive. On the basis of the second derivatives we would conclude that the functions were constrained minima. This conclusion is incorrect for the first example, since the constrained optimum was a maximum.

The Lagrangian multiplier provides an indication of the effect of a unit change in the constant in the constraining equation on the objective function. From the Lagrangian expression in (4.11), note that the expression $\lambda g(x, y)$ is added to the objective function to obtain the constrained optimum. Since $g(x, y)$ equals 0, the product of $\lambda g(x, y)$ is 0. If, however, we alter the constraining equation to allow $g(x, y)$ to assume an incrementally negative or positive value, then, provided λ is not 0, $g(x, y)$ is no longer 0. This term now adds to or subtracts from the objective function. The effect of $\lambda g(x, y)$ depends upon the magnitude of λ and the signs of both λ and $g(x, y)$.

To illustrate, assume that the constant in the constraining equation in the first example problem is increased by 1. The constraint is now $x + y = 101$, and $g(x, y) = x + y - 101$. For $x = 50$, $y = 50$, and $\lambda = -250$, $g(x, y) = -1$ and $\lambda g(x, y) = +250$. The effect of increasing the constant in the constraint by 1 results in the addition of a positive term to the constrained optimum. Similarly, if the constant in the constraining equation had been reduced by 1, the constrained optimum would be reduced.

The objective function is not linear. Therefore, the exact effect upon the objective function of a unit or incremental change in the constraint is not specified by the multiplier. The multiplier does, however, show the approximate effect of a one-unit change on the constrained optimum. This is shown by the following example.

Example. Determine the effect of increasing the constant in the constraint of the preceding problem to 101.

$$z = x^2 + 3xy + y^2,$$

subject to

$$x + y = 101.$$

$$F(x, y, \lambda) = x^2 + 3xy + y^2 + \lambda(x + y - 101).$$

Solving for the critical value gives

$$x^* = 50.5, \quad y^* = 50.5, \quad \text{and} \quad \lambda^* = -252.5$$

and

$$F(50.5, 50.5, -252.5) = 12{,}751.25.$$

The constrained optimum for $g(x, y) = x + y - 101$ exceeds the constrained optimum for $g(x, y) = x + y - 100$ by 251.25. This is approximately equal to $\lambda g(x, y) = -250(-1) = 250$.

Interpretation of the Lagrangian multiplier can be summarized as follows. The magnitude of the Lagrangian multiplier indicates the approximate change in the objective function for a unit change in the constant in the constraint. If λ is positive, the constrained optimum will increase if the constant in the constraint is decreased and decrease if the constant in the constraint is increased. If λ is negative, the constrained optimum will increase if the constant in the constraint is increased and decrease if the constant in the constraint is decreased.

Example. Determine the constrained optimum and the approximate effect of a unit change in the constraining equation for the function

$$z = 3x^2 - 6xy + y^2$$

subject to the constraint that

$$2x + y = 150.$$

The Lagrangian expression and the partial derivatives are

$$F(x, y, \lambda) = 3x^2 - 6xy + y^2 + \lambda(2x + y - 150)$$

$$\frac{\delta F}{\delta x} = 6x - 6y + 2\lambda = 0$$

$$\frac{\delta F}{\delta y} = -6x + 2y + \lambda = 0$$

$$\frac{\delta F}{\delta \lambda} = 2x + y - 150 = 0.$$

The critical values of x, y, and λ are $x^* = 39.5$, $y^* = 71$, and $\lambda^* = 95$. Substitution of these critical values into the Lagrangian expression gives a constrained optimum of $F(39.5, 71, 95) = -7105.25$. We can determine if this optimum is a maximum or minimum by application of the method established above. Using $\Delta x = 0.5$ and $\Delta y = 1.0$, we find that

$$F(39, 72, 95) = -7101,$$

and

$$F(40, 70, 95) = -7100.$$

The critical values of $x^* = 39.5$ and $y^* = 71$ yield a value, $F(39.5, 71, 95) = -7105.25$, which is smaller than at any adjacent point on the constraining plane. Consequently, the function reaches a constrained minimum at the critical points.

The method of determining constrained optima by Lagrangian multipliers can be expanded to include more than one constraint. For the general case of m variables with n constraints, a multiplier is introduced for each of the n constraints. The $(m + n)$ partial derivatives of the Lagrangian expression are equated to 0 and solved simultaneously for the m critical values and the n Lagrangian multipliers. This procedure is illustrated in the following example for three variables and two constraints.

Example. Determine the critical points of the function

$$z = f(x, y, w) = 2xy - 3w^2$$

subject to

$$x + y + w = 15 \quad \text{and} \quad x - w = 4.$$

Since we have two constraining equations, we must use two Lagrangian multipliers. The method of solution involves forming the Lagrangian expression and equating the partial derivatives with 0.

$$F(x, y, w, \lambda_1\lambda_2) = 2xy - 3w^2 + \lambda_1(x + y + w - 15) + \lambda_2(x - w - 4).$$

$$\frac{\delta F}{\delta x} = 2y + \lambda_1 + \lambda_2 = 0$$

$$\frac{\delta F}{\delta y} = 2x + \lambda_1 = 0$$

$$\frac{\delta F}{\delta w} = -6w + \lambda_1 - \lambda_2 = 0$$

$$\frac{\delta F}{\delta \lambda_1} = x + y + w = 15$$

$$\frac{\delta F}{\delta \lambda_2} = x - w = 4.$$

Solving the five equations simultaneously gives

$$x^* = 4.43, \quad y^* = 10.14, \quad w^* = 0.43, \quad \lambda_1 = -8.86, \quad \lambda_2 = -11.42.$$

The Lagrangian multipliers have the same signs, λ_1 and λ_2 being negative.

This indicates that if the constant in the first constraining equation is increased, the objective function will increase. Similarly, if the constant in the second constraining equation is increased, the objective function will also increase. The value of the function at the constrained optimum is

$$F(4.43, 10.14, 0.43, -8.86, -11.42) = 89.5.$$

The procedure for determining if the optimum is a constrained maximum or minimum is conceptually the same as expressed above. Given the function $F(x, y, w)$ and the constraining equations $g(x, y, w)$ and $h(x, y, w)$, we investigate points adjacent to the critical values of x, y, and w. For the case of three independent variables, 12 adjacent points must be evaluated. This set of 12 points is

$$\begin{aligned} S = [&(x + \Delta x, y - \Delta y, w), (x + \Delta x, y - \Delta y, w + \Delta w), \\ &(x + \Delta x, y - \Delta y, w - \Delta w), (x + \Delta x, y + \Delta y, w - \Delta w), \\ &(x - \Delta x, y + \Delta y, w), (x - \Delta x, y + \Delta y, w + \Delta w), \\ &(x - \Delta x, y + \Delta y, w - \Delta w), (x - \Delta x, y - \Delta y, w - \Delta w), \\ &(x, y + \Delta y, w - \Delta w), (x, y - \Delta y, w + \Delta w), \\ &(x + \Delta x, y, w - \Delta w), (x - \Delta x, y, w + \Delta w)]. \end{aligned}$$

Each adjacent point must be selected so that the constraining equations are satisfied, i.e., $g(x, y, w) = 0$ and $h(x, y, w) = 0$. If these conditions are met, the function is a constrained minimum when the function evaluated at the critical point is less than it is when it is evaluated at any adjacent point. It is a constrained maximum when the reverse is the case.

The method of Lagrangian multipliers is useful in allocating scarce resources between alternative uses. The following two example problems illustrate the Lagrangian technique for optimizing revenue functions subject to budgetary constraints.

Example. Revenue of the Rinehart Distributing Company is related to advertising A and the quantity Q produced and sold according to the function

$$R = 520 - 5A^2 + 21A + 15QA - 4.5Q^2 + 15Q.$$

The budgetary constraint for advertising and production is

$$2Q + A = 10.$$

Determine the values of Q and A that maximize revenue subject to the budgetary constraint.

To determine the critical values of the variables, we form the Lagrangian expression

$$\begin{aligned} F(A, Q, \lambda) = 520 &- 5A^2 + 21A + 15QA - 4.5Q^2 \\ &+ 15Q + \lambda(2Q + A - 10). \end{aligned}$$

The partial derivatives of this expression are equated to 0.

$$\frac{\delta F}{\delta A} = -10A + 15Q + \lambda + 21 = 0$$

$$\frac{\delta F}{\delta Q} = 15A - 9Q + 2\lambda + 15 = 0$$

$$\frac{\delta F}{\delta \lambda} = A + 2Q - 10 = 0.$$

The three equations are solved simultaneously, giving

$$A^* = 4.08, \qquad Q^* = 2.96, \quad \text{and} \quad \lambda = -24.6.$$

The maximum revenue subject to the budgetary constraint is

$$F(4.08, 2.96, -24.6) = 708.8.$$

From the value of the Lagrangian multiplier, we see that a one-unit increase in the budgetary constraint, i.e.,

$$2Q + A = 11,$$

would result in approximately 24.6 units of additional revenue.

Example. The revenue function for the Corola Typewriter Company was given on p. 106 as

$$R = 8E + 5M + 2EM - E^2 - 2M^2 + 20,$$

where E represented the number of electric portables and M the number of manual portable typewriters. The optimum profit occurred when $E^* = 10{,}500$ and $M^* = 6500$. The maximum revenue was $R =$ \$78,250. Assume that capacity constraints limit production to 14,000 typewriters. Determine the optimum product mix based upon this constraint.

The constraint can be incorporated into the problem through the method of Lagrangian multipliers. The Lagrangian expression is

$$F(E, M, \lambda) = 8E + 5M + 2EM - E^2 - 2M^2 + 20 + \lambda(E + M - 14).$$

The partial derivatives of the Lagrangian expression are equated to 0.

$$\frac{\delta F}{\delta E} = 8 + 2M - 2E + \lambda = 0$$

$$\frac{\delta F}{\delta M} = 5 - 4M + 2E + \lambda = 0$$

$$\frac{\delta F}{\delta \lambda} = -14 + M + E = 0.$$

Solving the three equations simultaneously gives

$$M^* = 5.3, \qquad E^* = 8.7, \quad \text{and} \quad \lambda^* = -1.2.$$

The sales revenue resulting from 5300 manual typewriters and 8700 electric typewriters is $R = \$76{,}500$.

4.3.3 INEQUALITY RESTRICTIONS

The method of Lagrangian multipliers can be modified to incorporate constraints that take the form of inequalities rather than equalities. The problem now becomes that of determining the extreme points of the multivariate function $z = f(x, y)$ subject to the inequality $g(x, y) \leq 0$ or $g(x, y) \geq 0$. We shall show a relatively simple extension of the Lagrangian technique which provides a solution to the problem of optimizing an objective function subject to a single constraining inequality. The general problem of optimizing an objective function subject to n inequalities is not considered in this text.

In the problem of optimizing an objective function subject to a constraining inequality, two cases must be considered. First, the constraining equality may act as an upper or lower bound on the function. Consider the objective function $z = x^2 + 3xy + y^2$ and the constraining inequality $x + y \leq 100$. In this problem, it is obvious that $x + y \leq 100$ acts as an upper bound on the values of x and y. Were it not for the constraining inequality, the function would reach a maximum at infinity.

The constraining inequality need not, however, act as an upper or lower bound on the function. As a second case, consider the function

$$z = -4x^2 + 4xy - 2y^2 + 16x - 12y.$$

The critical values of x and y were determined on p. 103 to be $x^* = 1$ and $y^* = -2$. If this problem were modified by the addition of the constraining inequality $x + y \leq 10$, the solution to the problem is not changed. Consequently, the constraining inequality does not act as an upper or lower bound on the function in this case.

For either of the above cases, the method of optimizing a function subject to a constraining inequality is to assume that the constraining inequality is an equality; that is, assume $g(x, y) = 0$. The inequality is changed to an equality and the critical values and constrained optimum are obtained by using the method of Lagrangian multipliers. The sign of the Lagrangian multiplier is used to determine whether the constraint is actually limiting the optimum value of the objective function. The procedure is as follows:

For maximizing the objective function subject to the constraining inequality $g(x, y) \leq 0$:

1. If $\lambda > 0$, the restriction is not a limitation; we resolve the problem, ignoring the restriction, to obtain the optimum.

2. If $\lambda \leq 0$, the restriction acts as an upper bound, and the result obtained by assuming that $g(x, y) = 0$ is the constrained optimum.

For minimizing the objective function subject to $g(x, y) \leq 0$:

1. If $\lambda > 0$, the restriction is a limitation, and the constrained optimum is obtained by assuming that $g(x, y) = 0$.
2. If $\lambda \leq 0$, the restriction is not a limitation; we resolve the problem, ignoring the restriction, to obtain the optimum.

For maximizing the objective function subject to $g(x, y) \geq 0$:

1. If $\lambda > 0$, the restriction is a limitation, and the constrained optimum is obtained by assuming that $g(x, y) = 0$.
2. If $\lambda \leq 0$, the restriction is not a limitation; we resolve the problem, ignoring the restriction, to obtain the optimum.

For minimizing the objective function subject to $g(x, y) \geq 0$:

1. If $\lambda > 0$, the restriction is not a limitation; we resolve the problem, ignoring the restriction, to obtain the optimum.
2. If $\lambda \leq 0$, the restriction is a limitation, and the constrained optimum is obtained by assuming that $g(x, y) = 0$.

The following examples illustrate this technique.

Example. Determine the maximum of the function $z = x^2 + 3xy + y^2$ subject to

$$x + y \leq 100.$$

To determine the solution we treat the inequality as an equality and form the Lagrangian expression

$$F(x, y, \lambda) = x^2 + 3xy + y^2 + \lambda(x + y - 100).$$

The problem was solved on p. 122. The critical values are

$$x^* = 50, \qquad y^* = 50, \quad \text{and} \quad \lambda^* = -250.$$

Since the Lagrangian multiplier is negative, we conclude that the value of the function is limited by the constraint. This is also obvious from inspection, since z approaches infinity as x or y approaches infinity. The extreme value of the function thus occurs as x and y reach the limiting value of the inequality.

Example. Consider again the Corola Typewriter Company. The revenue function for Corola, expressed in terms of electric E and manual M typewriters, was

$$R = 8E + 5M + 2EM - E^2 - 2M^2 + 20.$$

This function was optimized without limitations on production on p. 106. The maximum revenue of \$78,250 occurred when $E^* = 10.5$ and $M^* = 6.5$.

On p. 128, we revised the problem by assuming that production capacity constraints limited production to $E + M = 14$. This problem was solved to give $E^* = 8.7$, $M^* = 5.3$, and $\lambda^* = -1.2$ with revenue of \$76,500. The negative coefficient of the Lagrangian multiplier shows that the maximum is limited by the equality $E + M = 14$.

Assume that we reformulate the problem again by incorporating the inequality that production be less than or equal to 20,000 units; i.e.,

$$E + M \leq 20.$$

Considering the inequality as an equality, we form the Lagrangian expression

$$F(E, M, \lambda) = 8E + 5M + 2EM - E^2 - 2M^2 + 20 + \lambda(E + M - 20).$$

The partial derivatives are equated to 0, and the resulting three equations are solved simultaneously to give

$$E^* = 12.3, \qquad M^* = 7.7, \quad \text{and} \quad \lambda^* = 1.2.$$

Total revenue for the constrained optimum is $R = \$76{,}400$. This represents a reduction of \$1850 from the unconstrained optimum. The positive value of λ indicates that the unconstrained maximum is greater than the constrained maximum. The solution to the unconstrained problem satisfies the inequality; that is, $10.5 + 6.5 \leq 20.0$.

Example. The operations research department of the Cost Plus Defense Corporation has established a function that describes profits from army and navy contracts. This function is

$$P = 600 - 4A^2 + 20N + 2AN - 6N^2 + 12A,$$

where A = army contracts in millions of dollars.
N = navy contracts in millions of dollars.
P = profits in thousands of dollars.

Company policy dictates that a minimum of \$5 million in contracts will be accepted during the coming year. Thus the constraining inequality is

$$A + N \geq 5.$$

Determine the values of A and N that maximize profits.

To obtain the solution, we form the Lagrangian expression

$$F(A, N, \lambda) = 600 - 4A^2 + 20N + 2AN - 6N^2 + 12A + \lambda(A + N - 5).$$

The partial derivatives of this expression are equated to 0:

$$\frac{\delta F}{\delta A} = -8A + 2N + 12 + \lambda = 0$$

$$\frac{\delta F}{\delta N} = 2A - 12N + 20 + \lambda = 0$$

$$\frac{\delta F}{\delta \lambda} = A + N - 5 = 0.$$

The three equations are solved simultaneously to give

$$A = 2.584, \quad N = 2.416, \quad \text{and} \quad \lambda = 3.834.$$

Profits are \$630,080.

Profits, subject to the constraint, are maximized when army contracts of \$2,584,000 and navy contracts of \$2,416,000 are obtained. Since λ is positive, we recognize that the constraint acts as a lower bound on the values of the variables. This can be shown by determining the optimum value of A and N without the constraint. The optimum value of P, if we assume no constraining inequality, is found by equating the partial derivatives $\delta F/\delta A$ and $\delta F/\delta N$ to 0 and solving the resulting two equations simultaneously. The solution of these two equations gives $A = 2$ and $N = 2$. The profit from this level of sales is \$656,000.

problems

1. Find the first and second derivatives and the cross partial derivative of the following multivariate functions.
   a. $f(x, y) = 2x + 3xy + 3y$
   b. $f(x, y) = 2x^2 + 3xy^2 - 5y^2 + y^3$
   c. $f(x, y) = (x^2 + y^2)^3$
   d. $f(x, y) = (2x + xy^2 + y^2)^3$
   e. $f(x, y) = (x + y)(x + y)^2$
   f. $f(x, y) = (2x + 3y^2)^2/(x^2 + y^2)$
   g. $f(x, y) = x/y$
   h. $f(x, y) = \ln (x^2 + xy + y^2)$
   i. $f(x, y) = \ln (2x + y)^2$
   j. $f(x, y) = x^2 \ln (x^2 + y^2)$
   k. $f(x, y) = (x + 2y) \ln (x + 3y)$
   l. $f(x, y) = e^{x+y}$
   m. $f(x, y) = e^{(x^2+y^2)}$
   n. $f(x, y) = 3^{2x}$
   o. $f(x, y) = e^e$

2. Find the first and second derivatives for the following functions.
   a. $f(x, y, z) = 2x + 3y + 4z$
   b. $f(x, y, z) = (x^2 + xy + y^2 + xyz + yz^2)$
   c. $f(x, y, z) = (x^3 + xyz + y^2 + z^4)^{1/2}$
   d. $f(x, y, z) = \ln (x + y + z)$
   e. $f(x, y, z) = x + yz + (1/xyz)$
   f. $f(x, y, z) = e^{(2x+y+z)}$
   g. $f(x, y, z) = e^{x^2+y^2+z^2}$
3. Find the maxima, minima, or saddle point for the following functions.
   a. $f(x, y) = 2x^2 - 4x + y^2 - 4y + 4$
   b. $f(x, y) = 3x^2 + 24x + 16y^2 - 32y - 10$
   c. $f(x, y) = -x^2 + y^2 + 10x - 15y - 5$
   d. $f(x, y) = -8x^2 - 3y^2 - 144x - 12y + 15$
   e. $f(x, y) = 2x^2 - 8xy - 8x + 12y^2 - 48y + 24$
   f. $f(x, y) = 2x^2 - 4x + 8y^2 + 80y + 50xy + 100$
   g. $f(x, y) = -12x^2 + 24x + 10xy + 4y^2 - 200y + 16$
   h. $f(x, y) = 8x^2 + 16x + 3y^2 - 12y - 24xy - 15$
   i. $f(x, y) = \ln (x^2 - 4x + 2y^2 - 16y)$
   j. $f(x, y) = \ln (2x^2 - 24x - 6y^2 + 12y)$
   k. $f(x, y) = e^{(x^2-8x-y^2-4y)}$
   l. $f(x, y) = e^{(4x^2+40x-3y^2+12y)}$
   m. $f(x, y) = 4^{(2x^2-8xy-8x+12y^2-48y)}$
4. Find the maxima and minima of the following functions subject to the constraining equations or inequalities.
   a. $f(x, y) = 5x^2 + 6xy - 3y^2 + 10$ and $x + 2y = 24$
   b. $f(x, y) = 12xy - 3y^2 - x^2$ and $x + y = 16$
   c. $f(x, y) = x^2 + 2y^2 - xy$ and $x + y = 8$
   d. $f(x, y) = 3x^2 + 4y^2 - xy$ and $2x + y = 21$
   e. $f(x, y) = -3x^2 - 4y^2 + 6xy$ and $3x + y = 19$
   f. Minimize $f(x, y) = 4x^2 + 5y^2 - 6y$ and $x + 2y \geq 20$.
   g. Maximize $f(x, y) = 10xy - 5x^2 - 7y^2 + 40x$ and $x + y \leq 12$.
   h. Minimize $f(x, y) = 12x^2 + 4y^2 - 8xy - 32x$ and $x + y \leq 2$.
5. The sales of the Ace Novelty Company are a function of price p, advertising a, and the number of salesmen n. The functional relationship is

$$S = (10{,}000 - 700p)n^{2/3}a^{1/2}.$$

   The current price is $5.00, advertising is $10,000, and 99 salesmen are employed. Using the partial derivatives, determine
   a. The effect of an additional $1 of advertising expenditure upon sales.

b. The effect of raising price by \$0.01 upon sales.
c. The effect of employing one additional salesman.

6. The cost of construction of a project depends upon the number of skilled workers x and unskilled workers y. If the cost is given by

$$c(x, y) = 4000 + 9x^3 - 72xy + 9y^2,$$

a. Determine the number of skilled and unskilled workers that results in minimum cost.
b. Determine the minimum cost.

7. The yearly profits of a small service organization, Bill Athy, Inc., are dependent upon the number of workers x and the number of units of advertising y, according to the function

$$p(x, y) = 412x + 806y - x^2 - 5y^2 - xy.$$

a. Determine the number of workers and the number of units in advertising that results in maximum profit.
b. Determine maximum profits.

8. Suppose that the quantity q sold of a certain item is a function of the price p and advertising a, given by

$$q = [10{,}000 + 200(1 - e^{-0.50a})]e^{-0.25p}.$$

Use the partial derivatives to determine the effect of
a. A price change from \$2.00 to \$2.01 when advertising is \$10.
b. An increase in advertising from \$10 to \$11 when price is \$2.00.

9. Assume in Problem 6 that union contracts require that each skilled laborer must direct five or more unskilled laborers. Determine the number of skilled and unskilled laborers that leads to minimum cost.

10. In Problem 7, assume that workers receive \$100 per week (\$5200 per year) and that advertising cost is \$100 per unit. Working capital is such that a maximum total expenditure of \$100,000 can be made on labor and advertising.
a. Determine the optimum allocation of these funds.
b. Determine profits based upon this allocation.

5

multivariate and exponential business models

This chapter introduces several important business and economic models that are based upon the calculus of multivariate and exponential functions. These include production models, the least squares model, financial models, growth models, and production functions.

5.1 production models

In the discussion of production and inventory models in Sec. 3.2, we illustrated the economic order quantity model and the production lot size model. These two basic models are now modified to permit shortages in inventory. It is assumed in the following two models that orders that cannot be filled from inventory are backlogged. However, since there is a certain inconvenience to the customer in being required to wait for delivery of his purchase, a cost of shortage is incorporated in the model. This cost of shortage could be in the form of a discount to the customer, such as is typical in purchasing from a catalogue. It also could be a penalty to the merchant, which represents his estimate of the economic value of the ill will created by being unable to fill a customer's order.

The notation used for the following two models is the same as that in Sec. 3.2, with the exception that we include the cost of shortage. The inventory costs and decision variables are represented as follows:

a = cost of holding one unit of inventory for one time period.
b = cost of shortage of one unit for one time period.
c = setup cost incurred when a new production run is started or inventory is purchased.

k = production rate, in units per time period.
r = rate of demand, in units per time period.
q = quantity produced or ordered at each setup.

5.1.1 ECONOMIC ORDER QUANTITY WITH SHORTAGES

In this model we assume that shortages are permitted and backlogged. We also assume that inventory is purchased and arrives a fixed number of days following the placing of the order. As an example, consider a retailer who purchases stock from wholesalers. His customers place an order that is either immediately filled or filled from the next order received from the wholesaler. His annual demand is $r = 5000$ per year. Holding cost is $a = \$10$ per unit per year. Shortage cost, which includes a penalty for customer dissatisfaction and possible loss of the customer, is $b = \$15$ per unit short per year. Order cost (or setup cost) is $c = \$300$ per order. Orders are received $u = 10$ days after the order is placed. Determine the quantity and production cycle. The model is shown in Fig. 5.1.

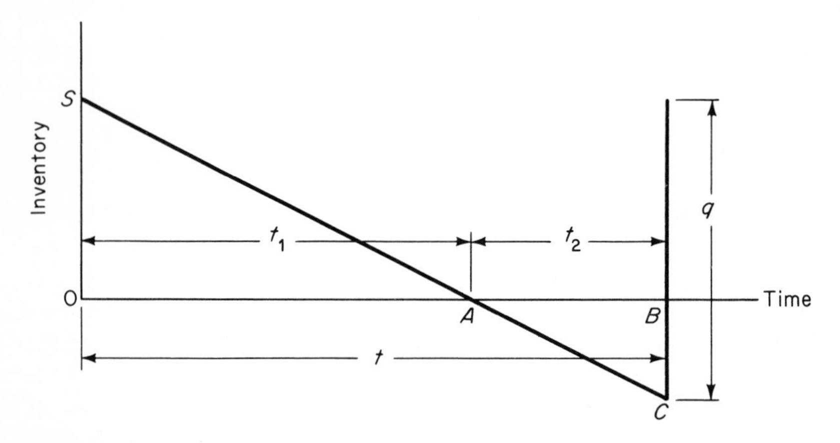

Figure 5.1

In Fig. 5.1, S represents the inventory level upon completion of one inventory cycle, t_1 represents the time during which orders are filled from inventory, and t_2 represents the time during which orders are backlogged. Both t_1 and t_2 represent fractions of the total period (i.e., fractions of the year) rather than of the inventory cycle.

From similar triangles, it can be seen that

$$\frac{t_1}{S} = \frac{t}{q}; \tag{5.1}$$

thus

$$t_1 = \frac{S \cdot t}{q} \tag{5.2}$$

and since

$$t_1 + t_2 = t, \tag{5.3}$$

we obtain

$$t_2 = t - t_1 = t - \frac{S \cdot t}{q} = (q - S)\,\frac{t}{q}\cdot \tag{5.4}$$

The cost of holding the inventory during one cycle is

$$C_c = \frac{a(St_1)}{2} = \frac{aS^2t}{2q}\cdot \tag{5.5}$$

The cost of shortage during the inventory cycle is

$$C_s = \frac{b(q - S)t_2}{2} = \frac{b(q - S)^2t}{2q}\cdot \tag{5.6}$$

Since the cost of placing an order is c, the inventory cost per cycle is

$$TC/\text{cycle} = c + \frac{aS^2t}{2q} + \frac{b(q - S)^2t}{2q}\cdot \tag{5.7}$$

The number of cycles per period is r/q. The total cost per period for inventory is thus

$$TC = \left[c + \frac{aS^2t}{2q} + \frac{b(q - S)^2t}{2q}\right]\frac{r}{q} + C_I, \tag{5.8}$$

and since r/q is equal to $1/t$, the expression reduces to

$$TC = \frac{cr}{q} + \frac{aS^2}{2q} + \frac{b(q - S)^2}{2q} + C_I. \tag{5.9}$$

In order to determine the values of q and S that minimize total cost, we equate the partial derivative of total cost with respect to q and the partial derivative of total cost with respect to S to 0. These two equations are solved simultaneously for q and S. Thus,

$$\frac{\delta(TC)}{\delta q} = \frac{-cr}{q^2} - \frac{aS^2}{2q^2} - \frac{b(q - S)^2}{2q^2} + \frac{b}{2q}\,[2(q - S)] = 0$$

$$\frac{\delta(TC)}{\delta S} = \frac{2aS}{2q} + \frac{2b(q - S)}{2q}\,(-1) = 0.$$

From the second equation we obtain

$$aS - bq + bS = 0,$$

or

$$S = \frac{bq}{(a+b)} \cdot \tag{5.10}$$

Substitution of S and solving for r in the first equation gives

$$2cr + \frac{a(bq)^2}{(a+b)^2} + b\left[q - \frac{bq}{a+b}\right]^2 - 2bq\left[q - \frac{bq}{a+b}\right] = 0$$

$$2cr + \frac{a(bq)^2}{(a+b)^2} + \frac{b(aq)^2}{(a+b)^2} - \frac{2bq(aq)}{(a+b)} = 0$$

$$2cr(a+b)^2 + abq^2(a+b) - 2abq^2(a+b) = 0$$

$$abq^2(a+b) - 2abq^2(a+b) = -2cr(a+b)^2$$

$$q^2 = \frac{-2cr(a+b)^2}{-ab(a+b)} = \frac{2cr(a+b)}{ab}.$$

$$q = \sqrt{\frac{2cr(a+b)}{ab}}. \tag{5.11}$$

Example. Determine the order quantity and the maximum level of inventory for the illustrative problem.

$$q = \sqrt{\frac{2(300)(5000)(10+15)}{10(15)}} = \sqrt{500{,}000} = 707 \text{ units},$$

$$S = \frac{bq}{a+b} = \frac{15(707)}{25} = 424 \text{ units}.$$

5.1.2 PRODUCTION LOT QUANTITY WITH SHORTAGES

This model describes the case in which more than one production run is made during the period under consideration. We assume that shortages are permitted at a cost of b per unit per time period and are backlogged. The production rate k is finite and greater than the usage rate r. A production and inventory usage cycle is shown in Fig. 5.2. It is important to note that t and t_i represent fractions of the time period rather than a fraction of the production cycle.

We shall first consider a single production-inventory cycle. Setup cost for the cycle is c. Shortage cost is the product of the average number of units short and the cost per unit short. Since OA and EF are equal,

$$C_s = \frac{EF(t_1 + t_4)b}{2}, \tag{5.12}$$

and since $EF = rt_4$, formula (5.12) can be expressed as

$$C_s = \frac{brt_4(t_1 + t_4)}{2}. \tag{5.13}$$

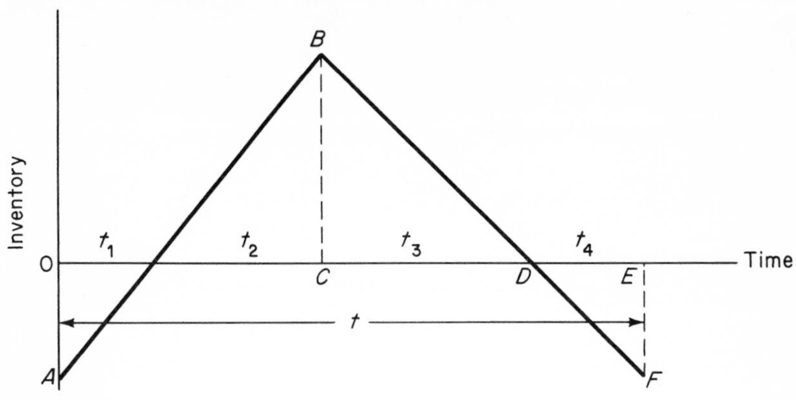

Figure 5.2

Inventory carrying cost during the cycle is given by

$$C_c = \frac{art_3(t_2 + t_3)}{2}. \tag{5.14}$$

The total cost of carrying the inventory, shortages, and the setup cost for one cycle is

$$TC/\text{cycle} = c + \tfrac{1}{2}[art_3(t_2 + t_3) + brt_4(t_1 + t_4)]. \tag{5.15}$$

The total cost for the period is determined by multiplying the cost per cycle by the number of cycles per period r/q and adding the cost of the product manufactured.

$$\begin{aligned} TC &= \frac{r}{2q}[art_3(t_2 + t_3) + brt_4(t_1 + t_4)] + \frac{cr}{q} + C_I \\ &= \frac{1}{2q}(art_3rt_2 + art_3rt_3 + brt_4rt_1 + brt_4rt_4 + 2cr) + C_I. \end{aligned} \tag{5.16}$$

Total cost can be expressed as a function of t_1 and t_2 by using the following equalities:

(i) $q = k(t_1 + t_2)$.
(ii) peak inventory $= t_2(k - r) = rt_3$.
(iii) peak shortage $= t_1(k - r) = rt_4$.

Substitution of these equalities yields

$$\begin{aligned} TC = \frac{1}{2q}[at_2^2(k - r)r + at_2^2(k - r)^2 + bt_1^2(k - r)r \\ + bt_1^2(k - r)^2 + 2cr] + C_I. \end{aligned} \tag{5.17}$$

By equating the partial derivatives of total cost with respect to t_1 and t_2 to 0, we find the optimal values of t_1 and t_2. These are

$$t_1 = \sqrt{\frac{2acr}{bk(k - r)(a + b)}}, \tag{5.18}$$

$$t_2 = \sqrt{\frac{2bcr}{ak(k - r)(a + b)}}. \tag{5.19}$$

By substitution for t_1 and t_2, we can obtain t and q.

$$t = \sqrt{\frac{2ck(a + b)}{arb(k - r)}}, \tag{5.20}$$

$$q = \sqrt{\frac{2crk(a + b)}{ab(k - r)}}. \tag{5.21}$$

Example. Assume that yearly demand is $r = 1000$ units and the production rate is $k = 6000$ units. Inventory holding costs are $a = \$10$ per unit per year, shortage costs are $b = \$15$ per unit per year, and setup costs are $c = \$200$ per setup. Determine q, t_1, t_2, t_3, t_4, and t.

$$q = \sqrt{\frac{2(200)(1000)(6000)(25)}{10(5000)(15)}} = \sqrt{80{,}000} = 283,$$

$$t_1 = \sqrt{\frac{2(10)(200)(1000)}{15(6000)(5000)(25)}} = \sqrt{0.0003555} = 0.01886 \text{ year},$$

$$t_2 = \sqrt{\frac{2(15)(200)(1000)}{10(6000)(5000)(25)}} = \sqrt{0.0008} = 0.02828 \text{ year},$$

$$t_3 = \frac{t_2(k - r)}{r} = \frac{0.02828(5000)}{1000} = 0.14140 \text{ year},$$

$$t_4 = \frac{t_1(k - r)}{r} = \frac{0.01886(5000)}{1000} = 0.09430 \text{ year},$$

and the inventory-production cycle is

$$t = t_1 + t_2 + t_3 + t_4 = 0.28284.$$

As a check on the calculations, we note that

$$t = \frac{q}{r} = \frac{283}{1000} = 0.283.$$

5.2 the method of least squares

An important application of differential calculus is the least squares model. The least squares model provides a method of fitting curves to data points. As an example, assume that an electric company is faced with the problem of forecasting sales of electricity for the next five years. Assume also that sales of electricity are dependent upon the number of homes in the area served by the company. From records for the past ten years, total sales and the total number of homes were obtained. These totals are plotted in Fig. 5.3.

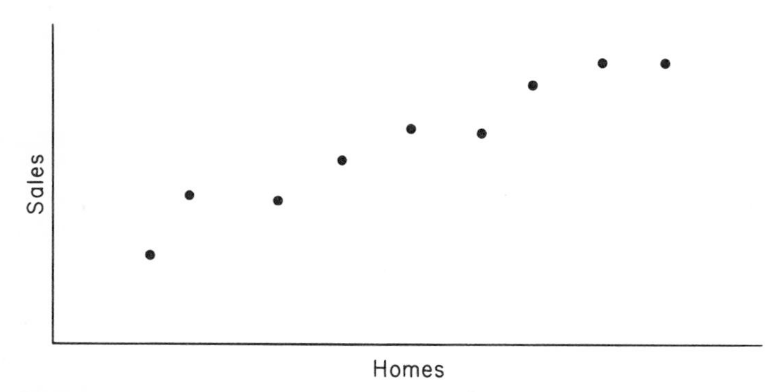

Figure 5.3

The plot of data in Fig. 5.3 is termed a *scatter diagram.* Each point is called a *data point.* The least squares model provides a method of establishing a function that describes the relationship between the variables, i.e., between the values on the vertical and horizontal axes. In terms of our example, this means establishing a functional relationship between electricity sales and the number of homes. The relationship would be used to forecast electricity sales on the basis of estimates of the number of homes in the area during the coming five-year period.

The scatter diagram in Fig. 5.3 shows that the relationship between the sales and the number of homes is approximately linear. Only two data points are necessary to establish a linear function. Since there are more than two data points, the analyst must either arbitrarily select two points that he believes are representative, or he must position the linear function through the data based upon some criterion of *best fit.* The method of least squares provides this criterion of best fit.

The method of least squares is commonly used to determine the relationship between variables in the regression model, the time series model, and econometric forecasting models. These subjects are beyond the design of this text. However, an understanding of the least squares model will be quite useful in the development of simple forecasting equations and in the study of these additional topics at a later date.

5.2.1 LINEAR FIT

The least squares model provides a method of establishing a curve through data points. The initial task is to determine the appropriate form of the curve, i.e., linear, exponential, quadratic, etc. A common method of determining the appropriate function is to plot the data points on a scatter diagram and determine by inspection the most appropriate form of the function. Fig. 5.4 provides an example in which the linear function applies.

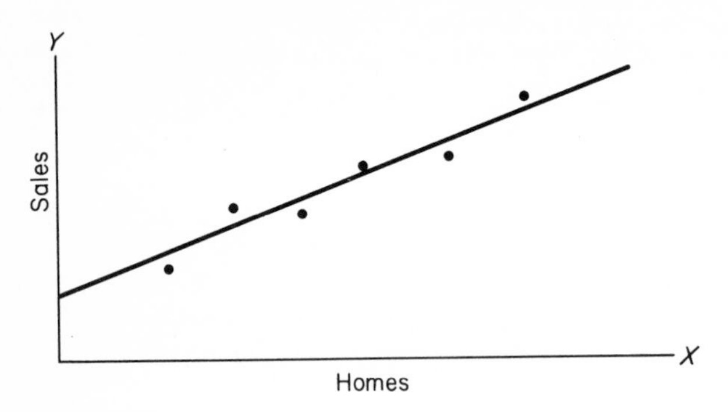

Figure 5.4

The general form of the linear function is

$$\tilde{y}_i = a + bx_i. \tag{5.22}$$

$\tilde{y}_i$ (read "y tilde") represents the value of y predicted for (based upon) a given value of x_i. The criterion of least squares is expressed mathematically as

$$\text{Minimize} \sum_{i=1}^{n} (y_i - \tilde{y}_i)^2, \tag{5.23}$$

where the symbol $\sum_{i=1}^{n}$, read "the summation as i is incremented in units of 1 from 1 to n," represents the summation of n terms.

Formula (5.23) states that the function is to be positioned on the scatter diagram in such a way that the summation of the squared deviations between the observed y values and the predicted y values is a minimum. We can substitute (5.22) in (5.23) and obtain

$$\text{Minimize} \sum_{i=1}^{n} [y_i - (a + bx_i)]^2. \tag{5.24}$$

Since we must determine a and b such that the sum of the squared deviations is a minimum, the function should be differentiated with respect to a, then with respect to b, and these partial derivatives should be equated with 0. Thus,

$$\frac{\delta F}{\delta a} = 2 \sum_{i=1}^{n} [y_i - (a + bx_i)]^1(-1) = 0,$$

and

$$\frac{\delta F}{\delta b} = 2 \sum_{i=1}^{n} [y_i - (a + bx_i)]^1(-x_i) = 0.$$

These two equations can be expressed as

$$\sum_{i=1}^{n} y_i = na + b \sum_{i=1}^{n} x_i, \tag{5.25}$$

and

$$\sum_{i=1}^{n} y_i x_i = a \sum_{i=1}^{n} x_i + b \sum_{i=1}^{n} x_i^2. \tag{5.26}$$

Equations (5.25) and (5.26) are termed the "normal equations." The normal equations are a set of two linear equations with unknown values a and b. Their simultaneous solution gives the values of a and b for the function that provides the best fit to the data according to the criterion of least squares.

The method of least squares is the most widely used technique for establishing the coefficients in the regression and econometric models. The least squares predicting function has certain characteristics that analysts find desirable. These characteristics are discussed in detail in statistical textbooks. The characteristics are

1. The least squares criterion provides the best fit to the data in the sense that the sum of the squared deviations, $\sum(y - \hat{y})^2$, of the observed values from the predicted values is a minimum.
2. The deviations above the line equal those below the line; that is, $\sum(y - \hat{y}) = 0$.
3. The linear function passes through the overall mean (average) of the data $(\bar{x}, \bar{y})$.
4. In those cases in which the data points are obtained by sampling from a larger population, the least squares estimates are "best estimates" of the population parameters in terms of the statistical properties of "unbiasedness and efficiency."

The following example illustrates use of the method of least squares in the development of a function for the forecasting of sales.

Example. Mr. Leonard Barker of the Finance Department of United Machinery Company wishes to determine the relationship between productivity and the profits of United Machinery. The productivity and profit data for six years are given in the following table. These data are plotted on the accompanying scatter diagram. Determine the predicting function for profit as a function of productivity.

	Profit (millions of dollars)	Productivity (output per man-week)
1963	3.1	145
1964	3.5	160
1965	3.6	185
1966	3.6	190
1967	3.8	200
1968	4.2	220

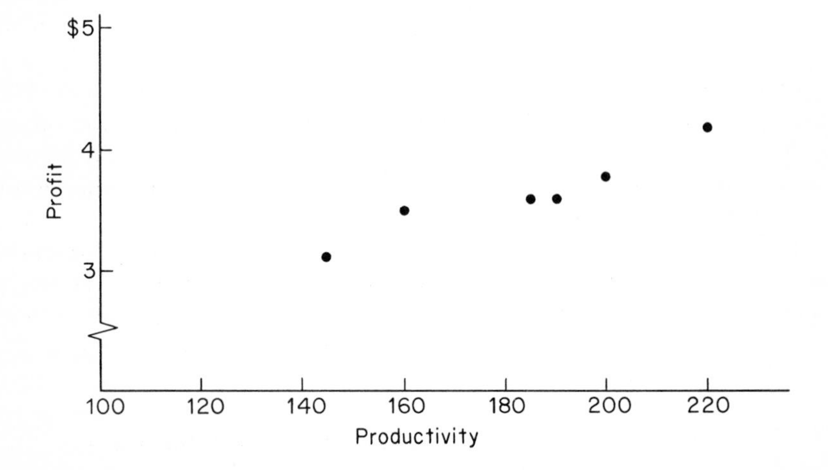

On the basis of the scatter diagram of profit versus productivity, we conclude that the relationship between the two variables is best described by a linear function. The general form of this function is given by (5.22). Calculation of the coefficients a and b is illustrated as follows.

x	y	$y \cdot x$	x^2
145	3.1	449.0	21,025
160	3.5	560.0	25,600
185	3.6	666.0	34,225
190	3.6	684.0	36,100
200	3.8	760.0	40,000
220	4.2	924.0	48,400
1100	21.8	4043.5	205,350

From these calculations $n = 6$, $\sum y = 21.8$, $\sum y \cdot x = 4043.5$, and $\sum x^2 = 205{,}350$. The values are substituted into the normal equations, and we have

$$21.8 = 6a + 1100b,$$
$$4043.5 = 1100a + 205{,}350b.$$

These two equations are solved simultaneously for a and b. The resulting least squares estimates of a and b are $a = 1.298$ and $b = 0.01274$. The predicting equation is

$$\tilde{y}_i = 1.298 + 0.01274x_i.$$

Mr. Barker, after consulting with the engineering staff of United Machinery, prepared estimates of productivity for 1969 through 1973. If we assume that the past relationship between productivity and profits remains the same, we can estimate profits using the predicting function. These estimates are shown in the following table.

Year	*Productivity*	*Profits*
1969	230	$4.23
1970	240	4.36
1971	245	4.42
1972	250	4.48
1973	255	4.55

5.2.2 CURVILINEAR FIT

The least squares linear fit model can also be used to establish nonlinear functions through data points. The criterion of least squares again provides the basis for selection of the parameters of the curvilinear function. We position the curvilinear function in such a way that the sum of the squared deviations between the observed y values and the predicted y values is a minimum. The resulting function provides estimates of the y values if the values of x are given. These estimates are based upon the least squares model.

Two of the more commonly used curvilinear functions are the exponential and power functions. The exponential function was introduced in Chapter 4. The general form of the exponential function is given by (4.8) as

$$y = ke^{cx}. \tag{4.8}$$

To apply the linear fit model to the exponential function, we express (4.8) in terms of natural logarithms.

$$\ln(y) = \ln(k) + cx. \tag{5.27}$$

Equation (5.27) is linear when expressed in terms of $\ln(y)$ and $\ln(k)$. This is apparent from the fact that $\ln(k)$ and c are constants; $\ln(y)$ is the dependent variable, and x is the independent variable. The values of $\ln(k)$ and c can be determined by solving Eqs. (5.25) and (5.26) for $\ln(k)$ and b. In Eqs. (5.25) and (5.26), we replace y_i by $\ln(y_i)$, a by $\ln(k)$, and b by c. Simultaneous solution of the two equations gives $\ln(k)$ and c.

Example. Bob's Family Restaurants operates restaurants throughout the southeastern United States. Sales for a five-year period are shown in the table and plotted in the scatter diagram below. If the growth trend that has been established over the five years were to continue for three more years, determine the predicting equation for sales and use this equation to predict sales for these three years.

SALES—BOB'S FAMILY RESTAURANTS

Year	*1965*	*1966*	*1967*	*1968*	*1969*
Sales	$200,000	$238,000	$290,000	$348,000	$430,000

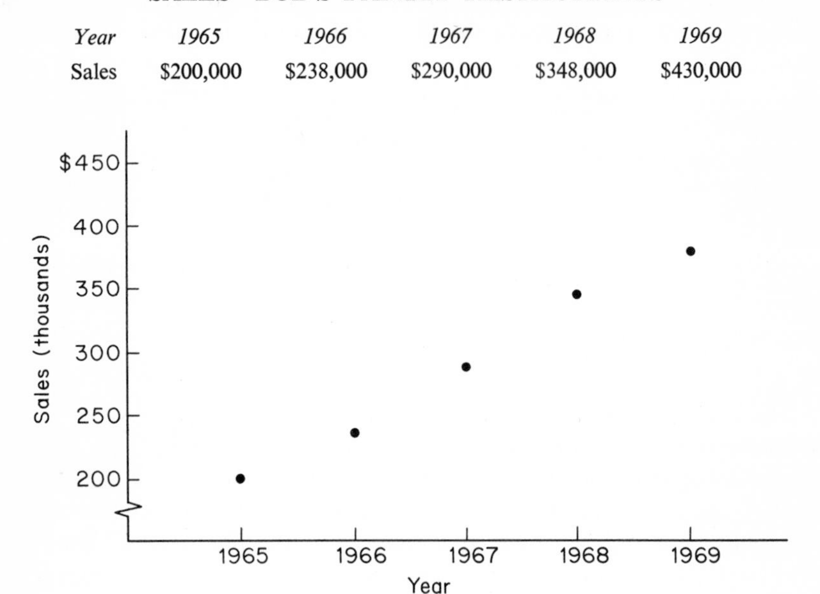

From the scatter diagram, we conclude that an exponential predicting equation best describes the growth in sales. Calculation of the coefficients k and c is illustrated below. To simplify the calculations, the five years are represented by $x = 1, 2, 3, 4, 5$.

x	y	$\ln(y)$	$x \ln(y)$	x^2
1	200	5.29832	5.29832	1
2	238	5.47227	10.94454	4
3	290	5.66988	17.00964	9
4	348	5.85220	23.40880	16
5	430	6.06379	30.31895	25
15		28.35646	86.97989	55

These values are substituted into the normal equations, giving

$$28.35646 = 5 \ln(k) + 15c,$$
$$86.97989 = 15 \ln(k) + 55c.$$

Solving the two equations simultaneously gives

$$\ln(k) = 5.09229 \quad \text{and} \quad c = 0.193.$$

The predicting function, written in terms of natural logarithms, is thus

$$\ln(y) = 5.09229 + 0.193x.$$

The sales forecast based upon the predicting function is $518,000 for 1970, $628,000 for 1971, and $762,000 for 1972.

The logarithmic function can be written as an exponential function by taking the antilogarithm of the function. This gives

$$y = 163e^{0.193x}.$$

The power function was introduced in Chapter 1. The general form of the power function is

$$y = ax^n. \tag{5.28}$$

This function is also linear when expressed in terms of logarithms.

$$\ln(y) = \ln(a) + n\ln(x). \tag{5.29}$$

We can apply the linear fit least squares model to this function by replacing y_i by $\ln(y_i)$, a by $\ln(a)$, b by n, and x_i by $\ln(x_i)$ in Eqs. (5.25) and (5.26). These two equations are solved simultaneously for $\ln(a)$ and n.

Example. Fit a power function of the form $y = ax^n$ to the sales data for Bob's Family Restaurants.

x	y	$\ln(x)$	$\ln(y)$	$\ln(x)\ln(y)$	$\ln(x)^2$
1	200	0	5.29832	0	0
2	238	0.69315	5.47227	3.79310	0.48046
3	290	1.09861	5.66988	6.22897	1.20694
4	348	1.38629	5.85220	8.11285	1.92180
5	430	1.60944	6.06379	9.75930	2.59030
		4.78749	28.35646	27.89422	6.19952

These values are substituted into the normal equations, giving

$$28.35646 = 5[\ln(a)] + 4.78749n,$$
$$27.89422 = 4.78749\ln(a) + 6.19952n.$$

Solving these two equations simultaneously gives $\ln(a) = 5.23094$ and $n = 0.45989$. The predicting equation written in terms of natural logarithms is $\ln(y) = 5.23094 + 0.45989[\ln(x)]$. The sales forecast based upon this predicting equation is $426,000 for 1970, $457,000 for 1971, and $487,000 for 1972.

Rewriting this predicting equation in the form of a power function gives

$$y = 187x^{0.45989}.$$

5.2.3 DETERMINING THE BEST CURVE

It is not always possible to determine by inspection the appropriate form of the predicting equation. In many cases, the data when plotted on a scatter diagram could be described by either the exponential or power function. It is also possible that the analyst is uncertain as to selection between a linear or power function, linear or exponential function, or perhaps between all three forms considered in this section.

The selection of the general form of the predicting equation should be based upon a careful study of the economic or physical relationship between the variables. It is possible, for example, that the data may appear to be best described by an exponential function. After careful analysis of the variables, however, the analyst may conclude that future values of the dependent variable are more likely to be linearly related to the independent variable. In this case, the linear function should be used instead of the exponential.

If the analyst believes that the selection of the general form of the predicting equation should be based solely upon the observed historical relationship between the variables, then that functional relationship which gives the minimum sum of the squared derivations is the appropriate function. The sum of the squared derivations is given by (5.23).

$$\sum_{i=1}^{n} (y_i - \tilde{y}_i)^2. \tag{5.23}$$

An alternative measure of the sum of the squared deviations is the *variance of the estimate.* This term is represented by s_e^2. The formula for the variance of the estimate is

$$s_e^2 = \frac{1}{n-2} \sum_{i=1}^{n} (y_i - \tilde{y}_i)^2. \tag{5.30}$$

The variance of the estimate is simply the sum of the squared deviations divided by $n - 2$. The functional relationship that gives the smallest variance of the estimate best describes the historical data.

Example. Determine the variance of the estimate for both the exponential and power function for the sales data for Bob's Family Restaurants.

Year	*Sales*	*Exponential estimate*	*Difference squared*	*Power estimate*	*Difference squared*
1965	200	197	9	187	169
1966	238	239	1	257	361
1967	290	290	0	310	400
1968	348	352	16	354	36
1969	430	428	4	392	1444
			30		2410

The variance of the estimate for the exponential function is

$$s_e^2 = \tfrac{30}{3} = 10.$$

The variance of the estimate for the power function is

$$s_e^2 = \tfrac{2410}{3} = 803.$$

Since the variance of the estimate for the exponential function is smaller than that for the power function, we conclude that the exponential function best describes the historical data.

5.3 mathematics of finance

The mathematics of finance is concerned with the time value of money. The rate of interest is one of the important variables in the business decision. Consequently, it is important to develop an understanding of the effect of interest rates on a sum of money over time. The formulas that describe these relationships are classified under the heading of mathematics of finance.

5.3.1 COMPOUND AMOUNT FORMULA

The formula for the amount, if we assume compounding n periods at a rate i per period, was given in Chapter 4. This formula is developed by recognizing that during the first period the amount of interest earned on P dollars is iP. The sum of the principal and interest at the end of the first period will be

$$P + iP = P(1 + i).$$

During the second period, interest will be earned on $P(1 + i)$ dollars. The sum of the interest and principal at the end of the second year would thus be

$$P(1 + i) + iP(1 + i) = P(1 + i)^2.$$

Continuing in this fashion, we find that the amount which results from compounding P dollars at interest rate i is given by

$$A = P(1 + i)^n. \tag{5.31}$$

This formula is termed the compound amount formula.

Example. Determine the interest and amount if \$1000 is invested at 6 percent for 8 years.

$$\begin{aligned} A &= 1000(1.06)^8 \\ \log A &= \log(1000) + 8 \log(1.06) \\ \log A &= 3.0000 + 8(0.0253) \\ \log A &= 3.2024 \\ A &= \$1593. \end{aligned}$$

The interest during the eight-year period totaled \$593, and the amount at the end of the eight-year period is \$1593.

Example. An individual has the option of investing \$5000 in a time deposit account that returns 5.50 percent interest or in mutual funds. The fee for purchasing the mutual funds is a one-time fee of $8\frac{1}{2}$ percent paid at the time of purchase of the fund. If the investment has a five-year duration, what rate of return must the fund yield to match that of the time deposit?

The amount at the end of the five-year period from the time deposit is

$$A = 5000(1.055)^5 = \$6535.$$

The principal invested in the mutual fund is

$$P = 5000 - 0.085(5000) = \$4575.$$

The interest rate is determined by solving the compound amount formula for i.

$$\begin{aligned} 6535 &= 4575(1 + i)^5 \\ \log(6535) &= \log(4575) + 5 \log(1 + i) \\ 3.81525 - 3.66040 &= 5 \log(1 + i) \\ 0.15485 &= 5 \log(1 + i) \\ \log(1 + i) &= \frac{0.15485}{5} = 0.03097 \\ 1 + i &= 1.072 \\ i &= 7.2\%. \end{aligned}$$

5.3.2 PRESENT VALUE

The present value of an amount of money received n periods in the future with interest of i per period can be determined by solving (5.31) for P.

$$P = A(1 + i)^{-n}. \tag{5.32}$$

In the present value formula, an amount A received n years from now has a present value of P. If we assume a positive interest rate, P will be less than A. P can be invested at rate i and will increase according to (5.31) to equal A. Consequently, P is termed the present value of A.

Example. Mr. Smith has an obligation of \$500 due five years from now. If interest is assumed to be 7 percent and is compounded yearly, what is the present value of the obligation?

$$\begin{aligned} P &= 500(1 + 0.07)^{-5} = 500(1.07)^{-5} \\ \log P &= \log 500 - 5 \log (1.07) \\ \log P &= 2.6990 - 5(0.0294) \\ \log P &= 2.5520 \\ P &= \$356. \end{aligned}$$

Example. A bond matures in 10 years and returns its holder \$1000. The bond is currently selling at \$800 and paying annually \$40 in interest payments. Determine the rate of interest received by the bondholder.

If the bond was purchased for \$800, the interest from the \$40 annual payment is

$$i_1 = \frac{\$40}{\$800} = 5\%.$$

The interest from the appreciation of the bond as it approaches maturity is

$$\begin{aligned} 1000 &= 800(1 + i)^{10} \\ \log (1.25) &= 10 \log (1 + i) \\ \frac{0.0969}{10} &= \log (1 + i) \\ \log (1 + i) &= 0.00969 \\ 1 + i &= 1.022 \\ i_2 &= 2.2\%. \end{aligned}$$

The total interest is, therefore, 7.2 percent.

The calculation of the amount and the present value has been made in the preceding examples by using logarithms. These calculations can be simplified through the use of tables of the compound interest and present value factors. A table of compound interest factors gives the value of the expression $(1 + i)^n$ for various values of i and n. Table I in the Appendix gives these factors. A table of present value factors gives the value of the expression $(1 + i)^{-n}$. These factors are given in Table II in the Appendix. The following examples illustrate the use of the compound interest and present value tables.

Example. If \$1500 is invested at 4 percent annual interest for 8 years, determine the amount of interest and principal.

$$A = \$1500(1.04)^8$$
$$A = \$1500(1.3686)$$
$$A = \$2053.$$

Example. Determine the present value of \$10,000 received 5 years from now, assuming 6 percent annual interest.

$$P = \$10{,}000(1.06)^{-5}$$
$$P = \$10{,}000(0.7473)$$
$$P = \$7473.$$

Example. If \$500 is deposited in a savings account paying 4 percent interest, compounded annually, what amount will be on deposit in the account at the end of 16 years?

$$A = \$500(1.04)^{16}$$
$$A = \$500(1.8730)$$
$$A = \$936.50.$$

Example. The purchasing power of the dollar declined at an average annual rate of 4 percent during 1968, 1969, and 1970. If we assume a continuation of this rate of decline, what will a 1970 dollar be worth in 1990?

$$P = \$1(1.04)^{-20}$$
$$P = \$1(0.4564)$$
$$P = \$0.4564 \text{ or } 45.64¢.$$

Example. The long-term growth rate of the United States gross national product has been 4 percent per year. If the gross national product in 1969 was \$930 billion and assuming a continuation in the growth rate, in what year will the gross national product reach \$2 trillion?

$$A = P(1.04)^n$$
$$2{,}000 = 930(1.04)^n$$
$$2.1505 = (1.04)^n.$$

From the table for amount at compound interest, we find that n is between 19 and 20 years. Consequently, we predict that the gross national product will surpass the \$2 trillion mark in 1989.

5.3.3 CONTINUOUS INTEREST

Interest is customarily expressed on an annual basis. If, for example, the interest rate is given as 5 percent per year, the annual interest rate is

$i = 5$ percent. The custom of stating interest on an annual basis is followed in practically all financial institutions.

The annual interest rate does not give the percentage of change in the amount invested during the year. If, for example, \$1 is invested at $i = 4$ percent and compounded quarterly, the percentage of change from the beginning to the end of the year will be 4.06 percent. This rate exceeds the 4 percent annual rate because the interest is added to the principal four times during the year rather than only at the end of the year. Interest is thus earned on interest during the year. This rate incorporates the effect of compounding n times during the year and is termed the *effective annual interest rate.*

The compound amount formula can be modified to include more than one compounding period per year. If i is the annual interest rate, t is the number of years, and n is the frequency per year of compounding, the compound interest formula becomes

$$A = P\left(1 + \frac{i}{n}\right)^{nt}, \tag{5.33}$$

where i/n is the interest rate per compounding period and nt is the number of periods.

Example. Determine the effective interest rate if the annual rate of 4 percent is compounded quarterly.

If we represent the effective annual rate by r, we can solve for the effective annual rate as follows:

$$\begin{aligned} (1 + r)^1 &= \left(1 + \frac{0.04}{4}\right)^4 \\ (1 + r) &= (1.01)^4 \\ 1 + r &= 1.0406 \\ r &= 0.0406 = 4.06\%. \end{aligned}$$

Example. Determine the present value of \$1000 received in 5 years if the annual interest rate is 6 percent and interest is compounded twice per year. Compare this to the present value if interest is compounded yearly.

Compounding twice a year gives

$$\begin{aligned} P &= 1000(1.03)^{-10} \\ P &= \$744. \end{aligned}$$

Compounding annually gives

$$\begin{aligned} P &= 1000(1.06)^{-5} \\ P &= \$747. \end{aligned}$$

Certain assets earn on a continuous basis throughout the year. As an example, capital equipment that is used in the manufacture of daily output contributes a continuous flow of earnings to the firm. The rate of return of this equipment is somewhat distorted if it is assumed that the return occurs only once each year. The actual rate of return should be calculated by incorporating the continuous flow of earnings into the formula.

The compound amount formula can be modified to incorporate continuous compounding. We must determine the effect of n approaching infinity on the compound amount formula. Thus,

$$A = \operatorname*{limit}_{n \to \infty} \left[P \left(1 + \frac{i}{n}\right)^{nt} \right],$$

which can be rewritten as

$$A = P \operatorname*{limit}_{n \to \infty} \left[\left(1 + \frac{1}{n/i}\right)^{n/i} \right]^{it}.$$

From formula (4.7), we recognize that the expression within the brackets equals e. Therefore, the formula for the amount for continuous compounding reduces to

$$A = Pe^{it}. \tag{5.34}$$

The present value of an amount A received t years hence if the principal is compounded continuously is determined by solving formula (5.34) for P.

$$P = Ae^{-it}. \tag{5.35}$$

Values of the term e^x and e^{-x} are given in Table V in the Appendix.

The use of Table V in determining the present value and amount if continuous compounding is assumed is illustrated by the following examples.

Example. Assume that \$10,000 is invested for 10 years at an annual interest rate of 6 percent. Compare the amount if the principal is compounded continuously and if it is compounded annually.

For continuous compounding:

$$A = 10{,}000e^{(0.06)10} = 10{,}000e^{0.6}$$
$$A = 10{,}000(1.822) = \$18{,}220.$$

For annual compounding:

$$A = 10{,}000(1.06)^{10}$$
$$A = 10{,}000(1.7908) = \$17{,}908.$$

The difference over the 10-year period between continuous compounding and annual compounding is \$312.

Example. One share of Miller Growth Fund, a mutual fund, was valued at \$10.00 on January 1 and \$11.00 on December 31. Determine the rate of return, assuming continuous appreciation throughout the year.

The method of determining i involves solving the continuous compounding formula for i. Thus

$$\begin{aligned} 11.00 &= 10.00e^{i(1)} \\ \ln(11.00) &= \ln(10.00) + i \\ i &= 2.39790 - 2.30259 \\ i &= 0.09531 = 9.53\%. \end{aligned}$$

Example. Determine the present value of a note of \$5000 that matures in 10 years. Assume continuous compounding and an annual interest rate of 7 percent.

$$\begin{aligned} P &= 5000e^{-(0.07)10} \\ P &= 5000e^{-0.7} \\ P &= 5000(0.497) \\ P &= \$2485. \end{aligned}$$

5.3.4 ANNUITIES

Annuities consist of a series of equal payments, each payment normally being made at the end of the period. Annuities are one of the most commonly used forms of financial agreements. Typical examples of annuities include mortgage payments, car payments, life insurance, retirement plans, sinking funds, and savings plans.

The amount of an annuity is the sum of the periodic payments and the interest earned from these payments. As an example, assume that an individual plans to save \$250 each quarter year for 5 years and that the savings will earn 4 percent interest compounded quarterly. The total of the savings and the interest earned on the savings can be calculated by using the formula for the amount of an annuity.

If we let i represent the interest rate per period, n the number of periods, p the annuity payment made at the end of each period, and A the sum of all the annuity payments and interest, we can calculate A as follows:

$$A = p + p(1 + i) + p(1 + i)^2 + \cdots + p(1 + i)^{n-1}. \qquad (5.36)$$

The final annuity payment is written first and earns no interest, the next to final payment earns interest for one period, and the first payment, which is made at the end of the first period, earns interest for $n - 1$ periods. This expression can be solved for A by a simple algebraic maneuver. This consists of multiplying the expression in (5.36) by $(1 + i)$. This gives

$$A(1 + i) = p(1 + i) + p(1 + i)^2 + p(1 + i)^3 + \cdots + p(1 + {}_i)^n. \qquad (5.37)$$

Subtracting (5.36) from (5.37) gives

$$A(1 + i) - A = p(1 + i)^n - p,$$

which when solved for A gives the formula for the amount of an annuity.

$$A = p\left[\frac{(1 + i)^n - 1}{i}\right]. \tag{5.38}$$

We can also determine the periodic payment necessary, when invested at interest rate i for n periods, to sum to the amount A. This is found by solving (5.38) for p.

$$p = A\left[\frac{i}{(1 + i)^n - 1}\right]. \tag{5.39}$$

The use of these formulas is illustrated by the following examples. In these examples, Table 1 in the Appendix is used to determine the value of $(1 + i)^n$. The annuity factor is then calculated by simple arithmetic.

Example. Mr. Clark plans on investing \$1000 per year in a savings plan that earns 5 percent interest compounded annually. Determine the sum of the annuity payments and interest at the end of 10 years.

$$A = \$1000\left[\frac{(1.05)^{10} - 1}{0.05}\right]$$

$$A = \$1000\left[\frac{1.6289 - 1}{0.05}\right]$$

$$A = \$1000(12.58) = \$12{,}580.$$

Example. Sioux Falls Steel Company recently placed a \$5 million bond issue with a group of private investors. One of the requirements of the investors was that the firm establish a sinking fund that will retire the bonds in 15 years. If Sioux Falls Steel plans to invest the sinking fund payments in government bonds that earn 6 percent interest, what yearly payment is necessary to retire the \$5 million bond issue at the end of 15 years?

$$p = \$5{,}000{,}000\left[\frac{0.06}{(1.06)^{15} - 1}\right]$$

$$p = \$5{,}000{,}000\left[\frac{.06}{2.3966 - 1}\right]$$

$$p = \$5{,}000{,}000(0.04296) = \$214{,}800.$$

It is often necessary to calculate the value of receiving \$$p$ per period for n periods, assuming an interest rate of i. This sum is termed the present value of an annuity. As an example, one might be required to determine the present value of receiving \$1000 per year for 10 years if interest is 6 percent per year.

Since money has a value over time, the present value of \$1000 per year for 10 years will be less than the value of receiving the \$10,000 during the first year. The present value of this series of annuity payments will, of course, depend upon the interest rate.

The formula for the present value of an annuity is derived by summing the present values of each of the individual annuity payments. The present value of the first payment, made one period from now, is $p(1 + i)^{-1}$. The present value of the second payment, made two periods from the present, is $p(1 + i)^{-2}$. The present value of all payments is the geometric series

$$PV = p(1 + i)^{-1} + p(1 + i)^{-2} + \cdots + p(1 + i)^{-n}. \tag{5.40}$$

If all terms in (5.40) are multiplied by $(1 + i)$, we obtain

$$PV(1 + i) = p + p(1 + i)^{-1} + \cdots + p(1 + i)^{-(n-1)}. \tag{5.41}$$

Subtracting (5.40) from (5.41) gives:

$$PV(1 + i) - PV = p - p(1 + i)^{-n},$$

which, when solved for PV, gives

$$PV = p\left[\frac{1 - (1 + i)^{-n}}{i}\right]. \tag{5.42}$$

The periodic payment that has a present value of PV is given by solving (5.42) for p:

$$p = PV\left[\frac{i}{1 - (1 + i)^{-n}}\right]. \tag{5.43}$$

The value of $(1 + i)^{-n}$ is given in Table 2 in the Appendix. The use of the present value and periodic payment formulas is illustrated by the following examples.

Example. Determine the present value of receiving \$1000 per year for 10 years if interest is 6 percent per annum.

$$PV = \$1000\left[\frac{1 - (1.06)^{-10}}{0.06}\right]$$

$$PV = \$1000\left[\frac{1 - 0.5584}{0.06}\right]$$

$$PV = \$1000(7.3601) = \$7360.$$

Example. Mr. Miller is considering the purchase of a home. The loan balance is \$20,000 and the interest rate is 7 percent. Determine the yearly payments for a 25-year loan.

$$p = \$20{,}000\left[\frac{0.07}{1 - (1.07)^{-25}}\right]$$

$$p = \$20{,}000\left[\frac{0.07}{1 - 0.1842}\right]$$

$$p = \$20{,}000(0.08581) = \$1716.$$

Example. An individual purchases \$1940 of furniture, which is to be paid for in 12 monthly payments of \$176 per month. What is the annual rate of interest?

The rate of interest can be determined by solving the present value of an annuity formula for i.

$$1940 = 176\left[\frac{1 - (1 + i)^{-12}}{i}\right]$$

$$11.02(i) = 1 - (1 + i)^{-12}.$$

We now must solve for i by trial and error. This is accomplished by selecting alternative values of i and iterating until a satisfactory value of i is determined. Assume

$$i = 0.01;$$

then

$$11.02(0.01) = 1 - (1.01)^{-12}$$

$$0.1102 < 0.1126.$$

Assume

$$i = 0.0125;$$

then

$$11.02(0.0125) = 1 - (1.0125)^{-12}$$

$$0.1378 < 0.1385.$$

Assume

$$i = 0.0150;$$

then

$$11.02(0.0150) = 1 - (1.0150)^{-12}$$

$$0.1653 > 0.1636.$$

From these calculations, it is apparent that i is between 0.0125 and 0.0150. Further refinement yields a monthly rate of approximately 0.013 or 1.3 percent. This corresponds to a yearly rate of 15.6 percent.

5.4 functions defined over time

In this section we consider the problem of when to sell an asset whose value is changing over time. The objective is to maximize the present value of the asset. To illustrate the technique, consider the following example.

Example. The Beck Investment Company has extensive land holdings in Florida. They have estimated that the value of these holdings is given by the functional relationship:

$$V = \$1{,}000{,}000(1.50)^{\sqrt{t}}.$$

Assuming a cost of capital of r percent (on a continuous basis) and disregarding any costs of upkeep of the undeveloped land, determine the optimal period of holding the land.

The present value of the land is given by formula (5.35). Thus,

$$P = Ve^{-rt} = 1{,}000{,}000(1.50)^{\sqrt{t}}\, e^{-rt}.$$

This can be written in its logarithmic form as

$$\ln P = \ln\,(1{,}000{,}000) + \sqrt{t}\ln\,(1.50) - rt.$$

To determine the optimal value of t, we equate the derivative of the function with 0 and solve for t. Since the function is logarithmic, Rule 16 is used for the derivative of logarithmic functions.

$$\frac{1}{P}\frac{dP}{dt} = \frac{1}{2}\ln\,(1.50)t^{-1/2} - r = 0$$

$$\frac{dP}{dt} = P\left[\frac{1}{2}\ln\,(1.50)t^{-1/2} - r\right] = 0.$$

Since P is not equal to 0, $\frac{1}{2}\ln\,(1.50)t^{-1/2} - r = 0$. Thus,

$$t^{-1/2} = \frac{2r}{\ln\,(1.50)},$$

and

$$t = \left[\frac{\ln\,(1.50)}{2r}\right]^2.$$

If we assume that $r = 8$ percent, then the optimum holding period for the land is

$$t = \left[\frac{\ln\,(1.50)}{0.16}\right]^2 = \left[\frac{0.40547}{0.16}\right]^2 = 6.4 \text{ years.}$$

The present value of the land is

$$\begin{aligned}
\ln(P) &= \ln(1{,}000{,}000) + \sqrt{6.4}\ln(1.50) - 0.08(6.4)\\
\ln(P) &= \ln(1{,}000{,}000) + 2.53(0.40547) - 0.08(6.4)\\
\ln(P) &= \ln(1{,}000{,}000) + 1.0258 - 0.512 = \ln(1{,}000{,}000) + 0.51384\\
P &= \$1{,}000{,}000(1.672) = \$1{,}672{,}000.
\end{aligned}$$

The technique utilized in determining the optimum time span for holding the land in the Beck Investment Company problem can be applied to any similar problem. The procedure involves expressing the value of the function in terms of its present value and maximizing the present value of the function. This involves equating the derivative of the present value function with 0 and solving for the optimum time interval. As a second illustration, consider the following example.

Example. Coleman and Associates are engaged in the business of importing, ageing, and distributing premium table wines. The value of the wine increases over time according to the following function:

$$V = 4.00(1.35)^{t^{2/3}}.$$

The present value of the wine, if we assume continuous appreciation and interest rate r, is

$$P = 4.00(1.35)^{t^{2/3}} e^{-rt}.$$

The optimum value of P is determined by equating the derivative of the present value function with 0.

$$\ln P = \ln (4.00) + t^{2/3} \ln (1.35) - rt.$$

$$\frac{1}{P}\frac{dP}{dt} = \frac{2 \ln (1.35)}{3} t^{-1/3} - r = 0.$$

Solving for t gives

$$t^{1/3} = \frac{2 \ln (1.35)}{3r}$$

$$t = \left[\frac{2 \ln (1.35)}{3r}\right]^3.$$

If we assume $r = 10$ percent, the optimum time for ageing the wine is

$$t = \left[\frac{2(0.300)}{0.30}\right]^3 = (2.00)^3 = 8.0 \text{ years.}$$

The present value of the wine is

$$P = 4.00(1.35)^{(8.0)^{2/3}}[e^{-0.10(8.0)}]$$

$$P = 4.00(1.35)^4[e^{-0.8}] = \$5.98 \text{ per bottle.}$$

The price per bottle obtainable at the end of the 8-year holding period is

$$P = 4.00(1.35)^{t^{2/3}} = 4.00(1.35)^4 = \$13.35.$$

The wine will thus be worth \$13.35 per bottle at the end of the 8-year period. Assuming that the cost of capital is 10 percent and assuming continuous compounding, we find that the present value of the wine is \$5.98 per bottle.

5.4.1 GROWTH RATE OF FUNCTIONS

The rate of growth of functions defined over time is given by the ratio of the derivative of the function and the function. Thus,

$$r = \frac{f'(t)}{f(t)}. \tag{5.44}$$

The derivative of the function represents the change in the function for an incremental change in time, whereas the denominator represents the base value of the function at time t. This ratio of the change in the function divided by the value of the function gives the rate of growth.

The meaning of (5.44) can perhaps be clarified by first considering a function that grows periodically rather than continuously. The formula for the growth rate of a discrete function is

$$r = \frac{\dfrac{\Delta f(t)}{\Delta t}}{f(t)}. \tag{5.45}$$

As an example, assume that \$2 is received quarterly on an investment of \$100. The rate of interest, expressed on an annual basis, is

$$r = \frac{\dfrac{2}{1/4}}{100} = 0.08 = 8\%.$$

Formulas (5.44) and (5.45) are quite similar, (5.44) being the limiting value of (5.45) as Δt approaches 0. The growth rate of continuous functions thus is given by (5.44) and that for discrete functions by (5.45).

The growth rate formula for continuous functions can be illustrated by the formula for the amount at continuous interest. Our objective is to show by using (5.44) that the growth rate of (5.34) is i. Applying (5.44) to $A = Pe^{it}$ we obtain

$$r = \frac{f'(t)}{f(t)} = \frac{Pe^{it}(i)}{Pe^{it}} = i.$$

As shown, the ratio of the marginal function and total function gives the rate of interest or growth rate. This relationship is illustrated in the following examples.

Example. Determine the continuous growth rate for $A = P(1 + i)^t$.

$$r = \frac{f'(t)}{f(t)} = \frac{P(1 + i)^t \ln (1 + i)}{P(1 + i)^t} = \ln (1 + i).$$

These examples illustrate that the growth rates for both the amount if continuous compounding is assumed and the amount if periodic compounding is assumed are constant. The constant growth rate if continuous compounding is assumed is i, whereas the constant growth rate for yearly compounding is $\ln(1 + i)$. We shall now consider several examples of growth functions in which the growth rate is not constant.

Example. The Beck Investment Company found that the value of their land investments could be described by

$$V = 1{,}000{,}000(1.50)^{\sqrt{t}}.$$

Determine the growth rate of this function.

$$r = \frac{\frac{dV}{dt}}{V} = \frac{1{,}000{,}000(1.50)^{\sqrt{t}}(1/2)t^{-1/2}\ln(1.50)}{1{,}000{,}000(1.50)^{\sqrt{t}}}$$

$$r = \frac{\ln(1.50)}{2\sqrt{t}} = \frac{0.40547}{2\sqrt{t}}.$$

The growth rate for Beck Investments depends upon the value of t. For $t = 1$, the growth rate is 20.27 percent. When $t = 4$ the rate has fallen to 10.14 percent.

The exponential function defined by formula (4.4) as

$$f(x) = k(a)^{cx}$$

is the only function that has the property of constant rates of growth. The growth rate for this function based upon (5.44) is $r = c\ln(a)$. If the function is rewritten as

$$A = Pe^{it},$$

or alternatively as

$$A = P(1 + i)^t,$$

the growth rate is seen to be i and $\ln(1 + i)$, respectively. Since the variable t is not included in the expressions for the growth rates, we conclude that the growth rate is constant for exponential functions. The characteristic of constant growth rates is, of course, the characteristic that leads to widespread usage of the exponential function in business and economic applications. Growth rates are not constant for algebraic functions. This can be seen by applying formula (5.44) to these types of functions. We illustrate the procedure for determining growth rates of algebraic functions with the following examples.

Example. The sales of a new product are described by the following function:

$$S(t) = 10{,}000t^2.$$

Determine the growth rate of sales.

$$r = \frac{S'(t)}{S(t)} = \frac{20{,}000t}{10{,}000t^2} = \frac{2}{t}.$$

The growth rate for this simple algebraic function is dependent upon the independent variable. This characteristic of varying growth rates is also true for more complicated algebraic functions.

Example. Following an advertising campaign, the sales of a product increase and then begin to decline. The sales are forecast by the quadratic function

$$S(t) = 1000 + 100t - 10t^2,$$

where t represents months. Determine the growth rate at the conclusion of the campaign ($t = 0$) and for each of the nine months following the conclusion. Determine sales during each of these months. The growth rate is given by (5.44).

$$r = \frac{S'(t)}{S(t)} = \frac{100 - 20t}{1000 + 100t - 10t^2}.$$

The growth rate and the sales are given for $t = 0$ through $t = 9$ in the following table.

t	0	1	2	3	4	5	6	7	8	9
r	10%	7.34%	5.16%	3.30%	1.61%	0%	−1.61%	−3.30%	−5.16%	−7.34%
S	1000	1090	1160	1210	1240	1250	1240	1210	1160	1090

5.5 production functions

The production of most goods requires the use of several factors of production. Economists describe the relationship between these inputs to production and the output of the finished commodity with a production function. The inputs or factors of production include land, labor, capital, and materials. These factors of production are considered as independent variables in the production function. The dependent variable is the quantity of product. If the quantity z of a product is produced by using two factors of production in amounts x and y, respectively, then the production function is

$$z = f(x, y). \tag{5.46}$$

The effect of adding additional units of one input while the other remains constant is given by the partial derivative of z with respect to the varying input. The partial derivative $\delta z/\delta x$ is termed the *marginal productivity of x.*

The marginal productivity of x describes the change in output for an incremental change in the input, if it is assumed that the remaining variable is held constant. Similarly, the marginal product of y is given by $\delta z/\delta y$ and describes the effect of varying y with x remaining constant.

Most production functions are characterized by positive marginal productivities over a considerable range of the variable. Since the inputs are very seldom perfect substitutes for each other, the marginal productivities become 0 or even negative if continual increases in a single input are attempted while the other inputs are held constant. This characteristic of production functions is termed *diminishing marginal productivity.*

The following examples illustrate the concept of production functions and marginal productivity.

Example. The production function for a firm is given by

$$z = 2x^{1/2}y^{1/4},$$

where z represents output, x represents capital input, and y represents labor input. Determine the marginal products of capital and labor for $x = 100$ and $y = 81$.

The marginal product of capital is

$$\frac{\delta z}{\delta x} = x^{-1/2}y^{1/4}.$$

The marginal product of labor is

$$\frac{\delta z}{\delta y} = \frac{1}{2}x^{1/2}y^{-3/4}.$$

The marginal products for both factors are positive for all values of the variables x and y. The function illustrates diminishing marginal productivity, however, in that $\delta z/\delta x$ decreases as x increases and similarly $\delta z/\delta y$ decreases as y increases. For $x = 100$ and $y = 81$, the marginal product of x is

$$\frac{\delta z}{\delta x} = (100)^{-1/2}(81)^{1/4} = \frac{(81)^{1/4}}{(100)^{1/2}} = \frac{3}{10}\cdot$$

To illustrate the diminishing marginal product, note that the marginal product for $x = 121$ and $y = 81$ is

$$\frac{\delta z}{\delta x} = (121)^{-1/2}(81)^{1/4} = \frac{(81)^{1/4}}{(121)^{1/2}} = \frac{3}{11}\cdot$$

The marginal product of y for $x = 100$, $y = 81$ is

$$\frac{\delta z}{\delta y} = \frac{1}{2}(100)^{1/2}(81)^{-3/4} = \frac{1}{2}(10)(3)^{-3} = \frac{5}{27}\cdot$$

The marginal product of y also decreases as y increases. If, for example, y is allowed to increase to $y = 256$, then

$$\frac{\delta z}{\delta y} = \frac{1}{2}(100)^{1/2}(256)^{-3/4} = \frac{1}{2}(10)(4)^{-3} = \frac{5}{64}.$$

The total product for $x = 100$ and $y = 81$ is

$$z = 2(100)^{1/2}(81)^{1/4} = 2(10)(3) = 60.$$

This example illustrates diminishing marginal productivity for the inputs to the production process. The marginal productivity for each input is, however, positive for all values of x and y. It is also possible for the marginal productivity to be negative. This occurs when additional quantities of an input to the production process result in a decrease in the total product.

Example. The production function for a firm is given by

$$z = 10xy - x^2 - 2y^2.$$

Determine the marginal productivity of x and y.

The marginal productivity of the input x is

$$\frac{\delta z}{\delta x} = 10y - 2x.$$

The marginal productivity of y is

$$\frac{\delta z}{\delta y} = 10x - 4y.$$

The marginal productivity of both x and y decreases as the variable increases. Furthermore, the marginal productivity of x becomes negative for $x > 5y$, and the marginal productivity of y becomes negative for $y > 2.5x$.

This type of production function describes a product requiring two inputs, e.g., labor and machinery. The marginal productivity of labor is positive as long as the labor units are less than five times as large as the machinery units. When the number of labor units applied to the production process exceeds five times the number of machinery units, the total output begins to fall. The decrease in output results from the increasingly larger number of workers who attempt to operate a fixed quantity of machinery. Similarly, if the number of workers remains constant and the units of machinery are continually increased, total product begins to fall when the units of machinery employed in the production process exceed two-and-one-half times the number of labor units. This occurs because a fixed number of workers are attempting to operate an increasing quantity of machinery. Total product begins to decline when the workers can no longer operate this larger quantity of machinery efficiently.

problems

1. An automobile transmission shop advertises 48-hour automatic transmission exchange. In order to guarantee 48-hour turn-around time, the shop maintains a relatively large stock of remanufactured transmissions. The cost of storing a transmission is $25 per year. If the shop does not have the proper transmission in stock, they must immediately request one from their supplier and pay for a rush delivery. The shortage cost averages $30. The shop expects a demand for 500 of a common type of transmission during the year. The normal cost of ordering the transmissions is $200 per order. Orders are received 5 working days after the order is placed. Determine the economic order quantity, the maximum inventory of this particular transmission, and the inventory cycle.
2. A television sales store can afford to stock only a limited number of color television sets because of the high cost of the inventory. A certain store expects to sell 250 of a particular set during the year. The holding cost is $30 per set per year. If the set is out of stock, the customer is offered a discount for the inconvenience of waiting for delivery of the set from the factory. The discount averages $50 per set short per year. The cost of placing an order for the particular set is $100. Determine the economic order quantity and the maximum quantity of this type of set in inventory at any one time.
3. The yearly demand for a particular product is 2000 units, and the production rate for this product is 8000 units per year. Inventory holding costs are $10 per unit per year, shortage costs are $15 per unit short per year, and setup costs are $500. Determine q, t_1, t_2, t_3, t_4, t, and the maximum inventory.
4. Ace Manufacturing Company manufactures various types of small electric motors, which are sold to various other firms as components. The yearly demand for a particular motor is 3000 units, and the production rate is 12,000 units per year. Inventory holding costs are $5.00 and the shortage costs are $2.00 per unit short. Setup costs are $350. Determine the production lot size, the maximum inventory, and the length of the production cycle.
5. An automobile retailer must determine the optimum number of a particular model of a new car to hold in inventory. Sales of the model are forecast as 30 units during the month of interest. The holding cost of the car is $60 per month. The shortage cost, which includes a discount to the customer, is $150 per unit short, and the cost of placing an order is $300. Determine the number of this model of car that should be ordered, the maximum number of cars

in inventory during the month, and the portion of the month in which customers cannot be promised immediate delivery.

6. The data that follow represent the results of a study of the relationship between productivity of workers and scores on selected aptitude tests.

Employee no.	*Y* *Productivity*	*X* *Aptitude score*
1	41.5	110
2	43.0	125
3	42.5	140
4	44.0	150
5	44.5	165
6	45.2	170
7	46.0	190
8	47.5	200
9	49.0	210
10	48.0	215
Total	451.2	1675

Sum of squares:

$$\sum y^2 = 20{,}414, \qquad \sum x^2 = 292{,}375, \qquad \sum xy = 76{,}558.$$

a. Plot these data on a scatter diagram and verify that a linear function can be used to predict productivity based upon aptitude score.
b. Determine the slope and intercept of the predicting function, using the method of least squares.
c. Determine the variance of the estimate.

7. The number of revenue passengers flown by Trans World Airlines for the period from 1959 through 1968 is given in the following table.

Year	*Year no.*	*Revenue passengers (millions)*
1959	0	5.9
1960	1	5.8
1961	2	5.5
1962	3	5.9
1963	4	6.8
1964	5	8.2
1965	6	9.7
1966	7	9.9
1967	8	12.9
1968	9	13.9
Total	45	84.5

Sum of squares:

$$\sum y^2 = 285, \quad \sum x^2 = 798.7, \quad \sum xy = 458.5.$$

a. Plot these data on a scatter diagram and verify that an exponential function can be used to describe the growth in revenue passengers.
b. Determine the exponential function, using the method of least squares.

8. From 1940 to 1970 the price of eggs rose from $0.23 per dozen to $0.70. If this rate of price increase continues, what would be the price of one dozen eggs in 2000?
9. Harvey Smith received a spendable income of $84.50 per week in 1960. His spendable income has increased to $124.00 per week in 1970. In 1960 the consumer price index was 103.1. The index was 134.0 in 1970. Determine the annual rate of increase in real spendable income for Mr. Smith.
10. Gross national product in the United States increased from $284.8 billion in 1950 to $503.7 billion in 1960. Determine the growth rate during this period.
11. Consolidated Edison bonds traded on February 16, 1970, for $603.75. These bonds mature in 1981 for $1000. The bonds pay $30 per year interest. Determine the interest rate yielded by the bond if it is held to maturity.
12. American Telephone and Telegraph bonds traded on February 16, 1970, for $615. The bond matures in 1984 for $1000. The bonds pay $32.50 per year interest. Determine the interest rate yielded by the bond if it is held to maturity.
13. Great Southern Savings and Loan currently pays 5 percent interest compounded quarterly on passbook savings accounts. If the compounding period is changed from quarterly to daily, determine the difference in the effective annual rate of interest.
14. A share of Investors Mutual declined in value from $10 per share to $9 per share between July 1 and December 31, 1969. Assuming continuous interest, determine the effective annual rate of decline for this mutual fund.
15. An individual can borrow $1000 at 8 percent interest, or he can discount a note for $1000 at 7.5 percent interest. Determine the effective annual interest for both alternatives.
16. If $100 is deposited at the beginning of every year for 5 years and compounded semiannually at a rate of 6 percent per year, determine the principal at the end of the five-year period.
17. Determine the yearly payments for a $30,000 home loan amortized over 30 years at 8 percent interest.

6

integral calculus

We have discussed differential calculus and business and economic models based upon differential calculus in the first five chapters of this text. In the study of calculus, it is customary to consider differential calculus and integral calculus as two related branches of calculus. The relationship between differential and integral calculus is established in Sec. 6.2 of this chapter. Before we turn to this relationship, it is important to understand the concept and meaning of integral calculus. Consequently, our first task in this chapter is to present the meaning of integral calculus. After this concept is clearly established, we can then discuss the relationship between the two branches of calculus and methods of determining integrals.

6.1 the integral

The task of defining the integral is best accomplished by illustrating the procedure for determining the area of an irregular figure. Assume that we wish to determine the area in Fig. 6.1 between the horizontal axis and the function $y = f(x)$ between $x = a$ and $x = b$. There is no geometrical formula that gives the area for this irregularly shaped function. We can, however, determine the approximate area of the figure by subdividing the area into rectangles and calculating the sum of the areas of the rectangles. The area has been subdivided into rectangles in Fig. 6.2.

The area of each rectangle in Fig. 6.2 is determined by dividing the distance from a to b into small intervals and treating each interval as the base of a rectangle. If n rectangles with equal base are constructed, the base of each rectangle is $\Delta x = (b - a)/n$. The height of the rectangle is given by the value of the function at the midpoint of each rectangle, $f(x_i)$. The ap-

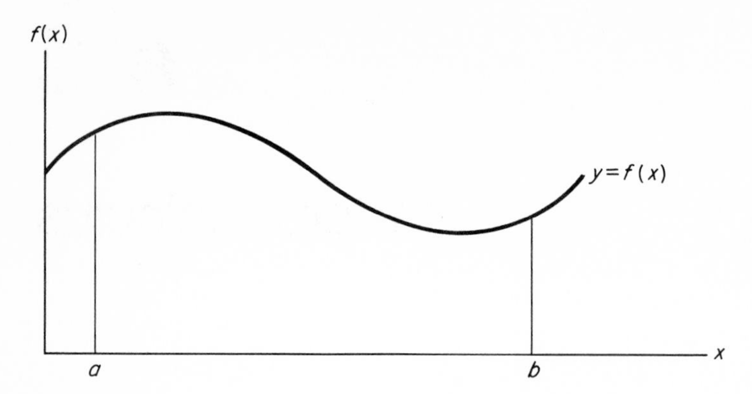

Figure 6.1

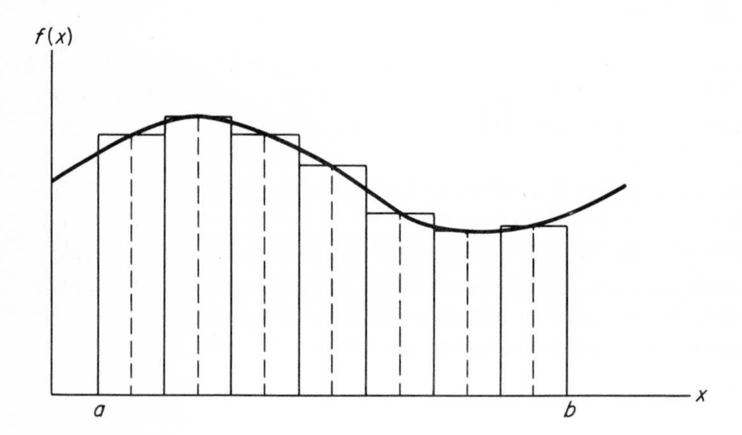

Figure 6.2

proximate area can now be found by summing the areas of the n rectangles. Since the area of a rectangle is given by the product of the base and height, $f(x_i)\,\Delta x$, the approximate area is given by

$$A \simeq f(x_1)\,\Delta x_n + f(x_2)\,\Delta x_n + \cdots + f(x_n)\,\Delta x_n, \tag{6.1}$$

where

$$\Delta x_n = \frac{b-a}{n}.$$

Using the Greek letter $\sum$ to represent the summation of a series of terms, we can rewrite (6.1) as

$$A \simeq \sum_{i=1}^{n} f(x_i)\,\Delta x_n, \tag{6.2}$$

where

$$\Delta x_n = \frac{b - a}{n}.$$

From Fig. 6.2, it can be seen that the accuracy of the approximation is dependent upon the number of rectangles in the interval from a to b. As the number of rectangles increases, the effect of the "gaps" between the function and the top of the rectangles on the area becomes of less importance. In other words, as Δx grows smaller or alternatively as n becomes larger, the approximation becomes more accurate. The area is given as the limit as Δx approaches 0 (i.e., as n approaches ∞) of the summation of the products of $f(x)$ and Δx. Expressed mathematically, the area is given by

$$A = \lim_{n \to \infty} \sum_{i=1}^{n} f(x_i)\,\Delta x_n, \tag{6.3}$$

where

$$\Delta x_n = \frac{b - a}{n}.$$

The magnitude of each term in the summation of (6.3) decreases as the number of terms increases. For instance, if Δx decreases from $\Delta x = 1.0$ to $\Delta x = 0.1$, then the magnitude of $f(x)\,\Delta x$ decreases by a factor of ten and the number of terms increases by a factor of ten. It can thus be seen that the area is given by the limiting value of the sum of a number of terms, when the number of terms increases infinitely as the value of each term approaches zero. This limiting value A is termed the *definite integral* of the function $f(x)$ for the interval $x = a$ to $x = b$.

The definite integral of a function $f(x)$ over the interval a to b is a numerical value that is associated with the function and the interval. More specifically, this numerical value is the limiting value of the sum of a number of terms when the number of terms increases infinitely as the numerical value of each term approaches zero. The symbolism commonly used for the definite integral of the function is

$$\int_a^b f(x)\,dx. \tag{6.4}$$

The summation sign is replaced by the elongated S or integral sign. a and b are, respectively, the lower and upper limits of integration. $f(x)$ represents the function for which the definite integral is being determined, and dx replaces Δx and is termed the *differential.*

The definite integral is used to determine quantities other than areas. We shall defer discussion of the use of the definite integral to determine quantities other than areas until Chapter 7. At this point, the student is expected to understand only the concept that the definite integral is the limiting value of the sum of a number of terms when the number of terms increases infinitely as the numerical value of each term approaches zero, and that the definite integral can be used to determine the area of an irregular shape.

6.2 relationship between integral and differential calculus

In order to determine the value of a definite integral, the reader must understand two additional concepts. These are the relationship between the *indefinite integral* and the derivative, and the *fundamental theorem* of calculus. We shall first show the relationship between the indefinite integral and the derivative. Sec. 6.3 introduces the fundamental theorem of calculus.

The indefinite integral and the derivative are related in the following manner. Assume that we want to find the indefinite integral of a function $f(x)$. The indefinite integral of the function $f(x)$ is another function $g(x)$ rather than a numerical value. The procedure for determining the indefinite integral of $f(x)$ is to find the function $g(x)$ such that the derivative of $g(x)$ is $f(x)$. If the derivative of $g(x)$ is $f(x)$, then $g(x)$ is the indefinite integral of $f(x)$. Thus,

$$\int f(x)\,dx = g(x), \tag{6.5}$$

provided that

$$g'(x) = f(x).$$

If $f(x)$ is thought of as the derivative of some function, then the indefinite integral of $f(x)$ is the function $g(x)$ such that $g'(x) = f(x)$. Integration is thus sometimes thought of as the reverse process of differentiation. For this reason, the term *antiderivative* is often used synonymously with the term *indefinite integral.* If we can find a function $g(x)$ and verify that the derivative of $g(x)$ equals $f(x)$, then we can conclude that $g(x)$ is the antiderivative or the indefinite integral of $f(x)$. These concepts are illustrated by the following examples.

Example. Determine $\int x\,dx$. Note that $f(x) = x$.

If $g(x) = \frac{1}{2}x^2 + c$, then $g'(x) = x$. By (6.5), we conclude that $\int x\,dx = \frac{1}{2}x^2 + c$, where c is any constant including 0.

Example. Find $\int (x^2 + 4x)\,dx$. In this example $f(x) = x^2 + 4x$. If $g(x) = \frac{1}{3}x^3 + 2x^2 + c$, then $g'(x) = x^2 + 4x$, and, therefore,

$$\int (x^2 + 4x)\,dx = \tfrac{1}{3}x^3 + 2x^2 + c.$$

Example. Determine $\int 2x\,dx$.

For $f(x) = 2x$,

$$g(x) = x^2 + c \quad \text{and} \quad \int 2x\,dx = x^2 + c.$$

The constant term in each of the preceding examples is necessary to illustrate that the indefinite integral of $f(x)$ is an entire family of functions. There are an infinite number of functions, each differing only by the value of the constant c, which have the derivative $f(x)$. In the preceding example, $g(x) = x^2 + c$ was the integral of $f(x) = 2x$. For any of the infinite number of possible values of the constant c, the derivative of $g(x)$ continues to be $f(x) = 2x$. In determining the indefinite integral, therefore, c is included as a part of the integral. This constant is termed the *constant of integration.* Fig. 6.3 illustrates several possible values of the constant of integration for the preceding example.

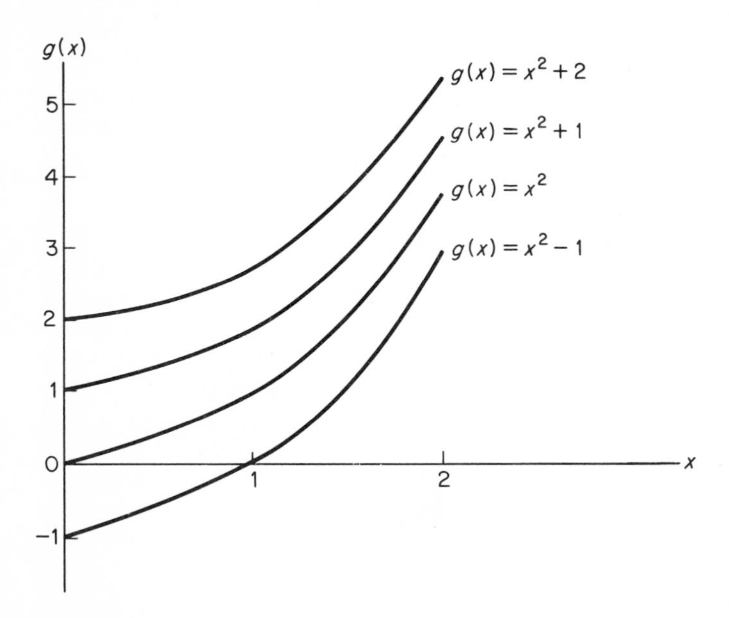

Figure 6.3

The constant of integration can only be determined if additional information is given. For example, if it is known that the value of the dependent variable for $x = 0$ is $g(0) = 1$ for the function sketched in Fig. 6.3, we know that the indefinite integral is $g(x) = x^2 + 1$. This additional information is called the *initial conditions.*

Example. Determine $\int x^{-1/2}\,dx$, given the initial condition that $g(4) = 6$.

$$g(x) = 2x^{1/2} + c \quad \text{and} \quad g(4) = 6;$$

thus

$$g(4) = 2(4)^{1/2} + c = 6 \quad \text{and} \quad c = 6 - 4 = 2.$$

Therefore,

$$g(x) = 2x^{1/2} + 2.$$

Example. Determine $\int (3x + 4)\,dx$, when $g(0) = 10$.

$$g(x) = \frac{3x^2}{2} + 4x + 10.$$

Methods of determining the indefinite integral of a function are discussed in Sec. 6.4.

6.3 fundamental theorem of calculus

The fundamental theorem of calculus relates the concepts of the definite integral and the indefinite integral. The definite integral of the function $f(x)$ between the limits $x = a$ and $x = b$ is a numerical value that can be determined from the indefinite integral. If we again represent the indefinite integral of the function $f(x)$ by $g(x)$, then the fundamental theorem of calculus states that

$$\int_a^b f(x)\,dx = g(b) - g(a). \tag{6.6}$$

This theorem states that the value of the definite integral of the function $f(x)$ between the limits a and b is given by the indefinite integral $g(x)$ evaluated at the upper limit of integration b minus the indefinite integral evaluated at the lower limit of integration a. Determining the definite integral thus merely involves evaluating the indefinite integral at $x = b$ and at $x = a$ and subtracting the two values. The constant of integration, which was required for the indefinite integral, is not included in the definite integral. This theorem is illustrated by the following examples.

Example. Using the methods of integral calculus, determine the area between the function $f(x) = x$ and the horizontal axis from $x = 0$ to $x = 5$.

The function $f(x) = x$ plots as shown below as a 45-degree triangle with base 5 and height 5. Using the formula for the area of a triangle $A = \frac{1}{2}bh$,

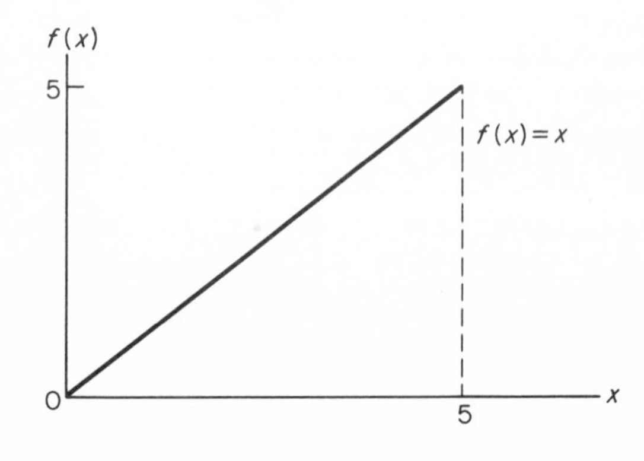

we observe that the area is 12.5. Using the concept of the definite integral, we have

$$A = \int_0^5 x\,dx = \frac{x^2}{2}\bigg|_0^5 = \frac{25}{2} - \frac{0}{2} = 12.5.$$

This example introduces the symbolism commonly employed in the evaluation of definite integrals. The integral of $\int x\,dx$ is $g(x) = x^2/2$. Since the limits of integration are $x = 0$ and $x = 5$, we determine $g(5)$ and $g(0)$.

$$g(5) = \frac{5^2}{2} = 12.5,$$

$$g(0) = \frac{0^2}{2} = 0,$$

and

$$g(5) - g(0) = 12.5.$$

An alternative way of expressing the same mathematical operation is to use the vertical line: $\Big|_a^b$. Thus we write

$$\int_0^5 x\,dx = \frac{x^2}{2}\bigg|_0^5.$$

The integral is then evaluated at the upper and lower limits, and the difference between these two values is the value of the definite integral.

This example can also be used to illustrate the reason that we do not include the constant of integration in the definite integral. Including the constant of integration for the integral evaluated at the upper limit gives

$$g(5) = 12.5 + c.$$

Similarly, evaluating the function at the lower limit gives

$$g(0) = 0 + c.$$

Subtracting the two gives

$$g(5) - g(0) = 12.5 + c - 0 - c = 12.5.$$

Since the constant is eliminated by the subtraction, it is unnecessary to include the constant. The symbolism and the concepts are further illustrated by the following examples:

Example. Evaluate the definite integral $\int_2^6 (x + 2)\,dx$.

$$\int_2^6 (x + 2)\,dx = \frac{x^2}{2} + 2x\bigg|_2^6 = \left(\frac{36}{2} + 12\right) - \left(\frac{4}{2} + 4\right) = 30 - 6 = 24.$$

Example. Evaluate the definite integral $\int_3^8 (2x + 4)\,dx$.

$$\int_3^8 (2x + 4)\,dx = x^2 + 4x\Big|_3^8 = (64 + 32) - (9 + 12) = 75.$$

A function must be continuous in the interval of integration in order to have a definite integral. This is illustrated by Fig. 6.4. Figure 6.4 shows a

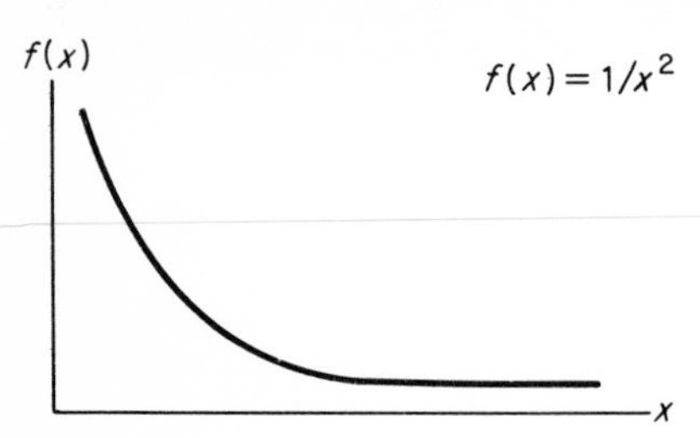

Figure 6.4

plot of the function $f(x) = 1/x^2$. This function cannot be evaluated for a lower limit of integration of $x = 0$. The function $f(x) = 1/x^2$ is not defined for $x = 0$ and is discontinuous at $x = 0$. The function does have a definite integral, however, for $x > 0$. This is illustrated by the following example.

Example. Evaluate the definite integral $\int_1^5 \frac{dx}{x^2}$.

$$\int_1^5 \frac{dx}{x^2} = \frac{-1}{x}\Big|_1^5 = \frac{-1}{5} - \frac{(-1)}{1} = \frac{4}{5}.$$

6.3.1 PROPERTIES OF DEFINITE INTEGRALS

Important properties of the definite integral can be illustrated with the aid of Fig. 6.5. This figure shows a continuous function $f(x)$ with values a, b, and c of x. One of the properties of the definite integral is that the definite integral of the function $f(x)$ with lower and upper limits of integra-

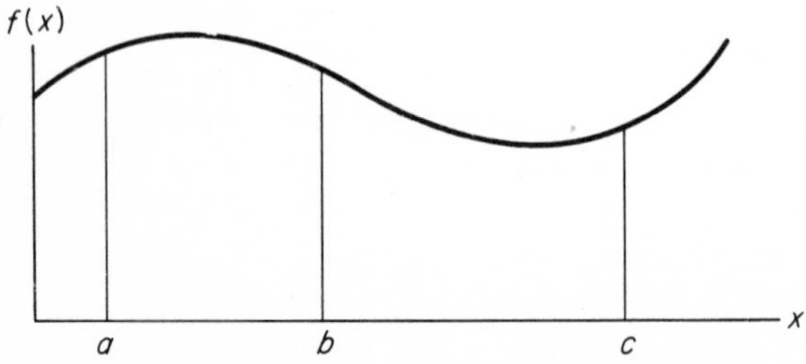

Figure 6.5

tion a and c, respectively, has the same value as the sum of the definite integrals of $f(x)$ with limits a and b and limits b and c. This is expressed as

$$\int_a^c f(x)\,dx = \int_a^b f(x)\,dx + \int_b^c f(x)\,dx. \tag{6.7}$$

This relation holds no matter what the relative values of a, b, and c.

A second property of the definite integral of the function $f(x)$ is that the value of the definite integral that has the same upper and lower limit is zero. Thus

$$\int_a^a f(x)\,dx = 0. \tag{6.8}$$

In the discussion of the definite integral, it was assumed that the lower limit of integration was numerically smaller than the upper limit of integration. It is also possible to determine the definite integral of a function over an interval in which the upper limit is numerically smaller than the lower limit. A third property of definite integrals is that the definite integral of the function $f(x)$ with limits of integration a to b is equal to the negative of the definite integral of the function with limits of integration b to a. This is expressed as

$$\int_a^b f(x)\,dx = -\int_b^a f(x)\,dx. \tag{6.9}$$

These properties are illustrated by the following examples.

Example. Show that $\int_1^5 x\,dx = \int_1^3 x\,dx + \int_3^5 x\,dx.$

$$\int_1^5 x\,dx = \left.\frac{x^2}{2}\right|_1^5 = \frac{25}{2} - \frac{1}{2} = 12,$$

$$\int_1^3 x\,dx = \left.\frac{x^2}{2}\right|_1^3 = \frac{9}{2} - \frac{1}{2} = 4,$$

and

$$\int_3^5 x\,dx = \left.\frac{x^2}{2}\right|_3^5 = \frac{25}{2} - \frac{9}{2} = 8.$$

Therefore,

$$\int_1^5 x\,dx = \int_1^3 x\,dx + \int_3^5 x\,dx.$$

Example. Show that $\int_3^6 x^2\,dx = -\int_6^3 x^2\,dx.$

$$\int_3^6 x^2\,dx = \left.\frac{x^3}{3}\right|_3^6 = \frac{(6)^3}{3} - \frac{(3)^3}{3} = 72 - 9 = 63,$$

$$\int_6^3 x^2\,dx = \frac{x^3}{3}\bigg|_6^3 = \frac{(3)^3}{3} - \frac{(6)^3}{3} = 9 - 72 = -63.$$

Therefore,

$$\int_3^6 x^2\,dx = -\int_6^3 x^2\,dx.$$

Example. Show that $\int_1^3 x\,dx = \int_1^5 x\,dx + \int_5^3 x\,dx.$

$$\int_1^3 x\,dx = \frac{x^2}{2}\bigg|_1^3 = \frac{9}{2} - \frac{1}{2} = 4$$

$$\int_1^5 x\,dx = \frac{x^2}{2}\bigg|_1^5 = \frac{25}{2} - \frac{1}{2} = 12$$

$$\int_5^3 x\,dx = \frac{x^2}{2}\bigg|_5^3 = \frac{9}{2} - \frac{25}{2} = -8.$$

Therefore, the definite integral is 4. This illustrates that (6.7) holds no matter what the relative values of a, b, and c.

Example. Evaluate $\int_2^2 x\,dx.$

$$\int_2^2 x\,dx = \frac{x^2}{2}\bigg|_2^2 = \frac{4}{2} - \frac{4}{2} = 0.$$

6.4 methods of integration

In comparing integral and differential calculus, most mathematicians would agree that the integration of functions is a more complicated process than differentiation of functions. Functions can be differentiated through application of a number of relatively straightforward rules. This is not true in determining the integrals of functions. Integration is much less straightforward and often requires considerable ingenuity. Certain functions can be integrated quite simply by applying rules of integration. Other functions require more complicated techniques, such as integration by substitution and integration by parts. Certain other functions can most readily be integrated through the use of tables of integrals. Still other rather ordinary-appearing functions cannot be integrated.

We shall discuss four methods of integration in this section. These are rules of integration, integration by substitution, integration by parts, and integration by use of tables of integrals. Additional methods of integration are available but are not included in this text.

6.4.1 RULES OF INTEGRATION

This section introduces rules for determining the integrals of certain functions and illustrates the rules through examples. When one is applying the rules for determining indefinite integrals, it is necessary to add the constant c to the integral. As stated earlier, this constant is determined from additional information concerning the integral. The constant may be nonzero or zero and is necessary, since the derivative $f(x)$ of the integral $g(x)$ does not retain any constant.

Rule 1. The integral of a constant is

$$\int k\,dx = kx + c.$$

For $k = 1$, the rule states that $\int dx = x + c$. The constant c is necessary, since the derivative of $kx + c$ is k.

Example. Determine $\int 10\,dx$.

$$\int 10\,dx = 10x + c.$$

Example. Determine $\int -20\,dx$.

$$\int -20\,dx = -20x + c.$$

Example. Determine $\int e\,dx$.

$$\int e\,dx = ex + c.$$

Rule 2. The integral of a variable to a constant power (power function) is

$$\int x^n\,dx = \frac{x^{n+1}}{n+1} + c \qquad \text{for } n \neq -1.$$

The student should carefully note that this rule applies for all power with the important exception of x^{-1}. A rule for determining the integral of x^{-1} is introduced on p. 182.

The integral of a power function is verified by noting that the derivative of $\dfrac{x^{n+1}}{n+1} + c$ is x^n.

Example. $\displaystyle\int x^3\,dx = \frac{x^4}{4} + c.$

Example. $\displaystyle\int x^{3/2}\,dx = \frac{2x^{5/2}}{5} + c.$

Example. $\displaystyle\int x^{-2}\,dx = -x^{-1} + c.$

Example. $\int x^k\,dx = \dfrac{x^{k+1}}{k+1} + c.$

Example. $\int x^{-3/4}\,dx = 4x^{1/4} + c.$

Rule 3. The integral of a constant times a function is the constant times the integral of the function.

$$\int kf(x)\,dx = k\int f(x)\,dx.$$

Example. $\int 12x^2\,dx = 12\int x^2\,dx = \dfrac{12x^3}{3} + c = 4x^3 + c.$

Example. $\int \tfrac{1}{4}x\,dx = \tfrac{1}{4}\int x\,dx = \dfrac{x^2}{8} + c.$

Example. $\int 6x^{-5}\,dx = 6\int x^{-5}\,dx = -\dfrac{6}{4}x^{-4} + c.$

Example. $\int -3x^{-1/2}\,dx = -3\int x^{-1/2}\,dx = -6x^{1/2} + c.$

Example. Evaluate the definite integral $\int_0^6 2x\,dx.$

$$\int_0^6 2x\,dx = x^2\Big|_0^6 = (36 - 0) = 36.$$

Rule 4. The integral of the sum or difference of two or more functions is the sum or difference of their integrals. That is,

$$\int [f(x) + g(x)]\,dx = \int f(x)\,dx + \int g(x)\,dx$$

and

$$\int [f(x) - g(x)]\,dx = \int f(x)\,dx - \int g(x)\,dx.$$

Example. $\int (x + 4)\,dx = \int x\,dx + \int 4\,dx = \dfrac{x^2}{2} + 4x + c.$

Example. $\int (x^2 - 3x + 6)\,dx = \int x^2\,dx - \int 3x\,dx + \int 6\,dx$

$$= \frac{x^3}{3} - \frac{3x^2}{2} + 6x + c.$$

Example. $\int (x^{-3} - x^{-2})\,dx = \int x^{-3}\,dx - \int x^{-2}\,dx$

$$= -\frac{x^{-2}}{2} + x^{-1} + c.$$

Example. Evaluate the definite integral $\int_{-1}^{5} (3x - 4)\, dx$.

$$\int_{-1}^{5} (3x - 4)\, dx = \int_{-1}^{5} 3x\, dx - \int_{-1}^{5} 4\, dx = \frac{3x^2}{2}\Big|_{-1}^{5} - 4x\Big|_{-1}^{5}$$

$$= \tfrac{3}{2}[(5)^2 - (-1)^2] - 4[5 - (-1)]$$

$$= \tfrac{3}{2}(25 - 1) - 4(5 + 1)$$

$$= 36 - 24 = 12.$$

Rule 5. The integral of the exponential function is

$$\int a^{kx}\, dx = \frac{a^{kx}}{k \ln (a)} + c.$$

Example. $\int 3^{2x}\, dx = \frac{3^{2x}}{2 \ln 3} + c.$

Example. $\int 10^{0.5x}\, dx = \frac{10^{0.5x}}{0.5 \ln 10} + c.$

Example. $\int e^{x}\, dx = \frac{e^{x}}{\ln e} + c = e^{x} + c$, since $\ln e = 1$.

Example. $\int 4e^{2x}\, dx = \frac{4e^{2x}}{2} + c = 2e^{2x} + c.$

Example. Evaluate the definite integral $\int_{0}^{1} 4e^{-4x}\, dx$.

$$\int_{0}^{1} 4e^{-4x}\, dx = -e^{-4x}\Big|_{0}^{1} = (-e^{-4}) - (-e^{0})$$

$$= -0.018 + e^{0} = -0.018 + 1 = 0.982.$$

Rule 6. The integral of the logarithmic function is

$$\int \log_a (kx)\, dx = x \log_a (kx) - x \log_a e + c.$$

This rule is verified by determining the derivative of the integral.

$$\frac{d}{dx}[x \log_a (kx) - x \log_a e + c]$$

$$= x\left(\frac{k}{kx}\right) \log_a e + \log_a (kx) - \log_a e = \log_a x.$$

If the function is $\ln (kx)$, the integral becomes

$$\int \ln (kx)\, dx = x \ln (kx) - x + c.$$

Example. Evaluate the definite integral $\int_1^4 \ln x\, dx$.

$$\int_1^4 \ln x\, dx = x \ln x - x\Big|_1^4 = (4 \ln 4 - 4) - (\ln 1 - 1)$$
$$= 4 \ln 4 - \ln 1 - 3 = 4(1.38629) - 0 - 3$$
$$= 2.54516.$$

Example. Determine $\int \ln (3x)\, dx$.

$$\int \ln (3x)\, dx = x \ln (3x) - x + c.$$

Rule 7. The integral of $f(x) = x^{-1}$ is

$$\int \frac{dx}{x} = \int x^{-1}\, dx = \ln x + c.$$

The rule applies when the variable x is raised to the constant power $n = -1$. For x raised to any power other than $n = -1$, Rule 2 rather than Rule 7 applies. Note that $f(x) = x^{-1}$ is equivalent to $f(x) = \frac{1}{x}$.

6.4.2 VARIABLE LIMITS OF INTEGRATION

The limits of integration (a, b) in the preceding examples have been assumed to have fixed numerical values. In certain problems, such as those often encountered in the study of probability, one of the limits of integration is a variable rather than a constant. As an illustration, let us consider the integral of $f(x)$ over the variable interval (a, x). The value of the definite integral is a function of the upper limit of integration. The definite integral is

$$F(x) = \int_a^x f(t)\, dt. \tag{6.10}$$

We have denoted the variable of integration as t rather than x to avoid confusion with the upper limit of integration.

Example. Determine the definite integral $F(x)$ for the function $f(x) = x$ for the variable limits of integration $(0, x)$.

This problem is the same as that discussed on p. 174 for the area of the triangle. The hypotenuse of the triangle is described by the function $f(x) = x$ and the base of the triangle is given by the interval $(0, x)$. The area of the triangle is a function of the length of the base $(0, x)$. For a variable base the area is

$$F(x) = \int_0^x t\, dt = \frac{t^2}{2}\Big|_0^x = \frac{x^2}{2}.$$

In the earlier example the base was $x = 5$.

$$F(5) = 12.5.$$

This agrees with the earlier calculations.

Example. Determine the definite integral $F(x)$ for the function $f(x) = x^2 - 2x$ for $(2, x)$. We replace x by t as the variable in the function to obtain

$$F(x) = \int_2^x (t^2 - 2t)\,dt = \frac{t^3}{3} - t^2 \Big|_2^x = \frac{x^3}{3} - x^2 - \frac{8}{3} + 4$$

$$F(x) = \tfrac{1}{3}(x^3 - 3x^2 + 4).$$

The relationship that exists between the function $F(x)$ and $f(x)$ is that the derivative of $F(x)$ is $f'(x)$. The function $F(x)$ defined by (6.10) has the derivative given by

$$F'(x) = f(x). \tag{6.11}$$

This can also be expressed as

$$\frac{d}{dx}\int_a^x f(t)\,dt = f(x).$$

This formula states that the derivative of a definite integral with respect to an upper limit of integration is equal to the function evaluated at the upper limit.

Example. Determine the derivative of $F(x) = \int_0^x t\,dt$.

$$F'(x) = \frac{d}{dx}\int_0^x t\,dt = x.$$

Example. Determine the derivative of $F(x) = \int_2^x (t^3 + 2t^2)\,dt$.

$$F'(x) = \frac{u}{dx}\int_2^x (t^3 + 2t^2)\,dt = x^3 + 2x^2.$$

6.4.3 INTEGRATION BY SUBSTITUTION

The rules of integration discussed in the preceding sections directly apply only to a limited number of types of functions. The use of the rules of integration can be extended by the technique of substitution. Certain functions can be integrated by substituting a variable for a function and integrating with respect to the substituted variable. If the function to be integrated is of the form

$$\int f(x)\,dx = \int h(g(x))g'(x)\,dx,$$

the integration can be performed by substitution. The function $h(g(x))$ is a composite function. The function $f(x)$ is equal to the product of the composite function $h(g(x))$ and the derivative of $g(x)$, $g'(x)$. The method of substitution involves first substituting the variable u for the function $g(x)$, i.e. $u = g(x)$. The function $h(g(x))$ now becomes $h(u)$. The next step is to determine the derivative of u with respect to x. This is

$$\frac{du}{dx} = g'(x).$$

In the study of differential calculus, du/dx was considered as a single symbol. In integral calculus we give separate meanings to du and dx, so that du/dx is now considered as a quotient: the differential of u divided by the differential of x. Since we are now considering du and dx as separate terms, we can write

$$du = g'(x)\,dx$$

or alternatively,

$$dx = \frac{du}{g'(x)},$$

provided

$$g'(x) \neq 0.$$

The substitution can now be made in the integral

$$\int f(x)\,dx = \int h(g(x))g'(x)\,dx.$$

We substitute $u = g(x)$ and $dx = \dfrac{du}{g'(x)}$ to obtain

$$\int h(g(x))g'(x)\,dx = \int h(u)\,du. \tag{6.12}$$

The integral of $h(u)$ is now determined.

The method of integration by substitution is illustrated by the following examples.

Example. Determine $\displaystyle\int \frac{2x\,dx}{(x^2 + 3)}$.

This integral can be expressed as $\int (x^2 + 3)^{-1} 2x\,dx$. If

$$g(x) = x^2 + 3,$$

then

$$g'(x) = 2x.$$

If we substitute u for $g(x)$, then

$$u = x^2 + 3$$

and

$$\frac{du}{dx} = 2x \quad \text{or} \quad du = 2x\,dx.$$

Substituting these values gives

$$\int u^{-1}\,du = \ln u + c.$$

By resubstituting $g(x)$ for u, we obtain

$$\int (x^2 + 3)^{-1} 2x\,dx = \ln (x^2 + 3) + c.$$

Example. Determine $\int x e^{x^2+9}\,dx$.

Let

$$g(x) = u = x^2 + 9.$$

The derivative of u with respect to x gives

$$\frac{du}{dx} = 2x,$$

or since we are considering du and dx as separate symbols

$$\frac{du}{2x} = dx.$$

Based upon the substitution, the integral becomes

$$\tfrac{1}{2}\int e^u\,du = \tfrac{1}{2}e^u + c.$$

By resubstituting $g(x)$ for u, we obtain

$$\int x e^{x^2+9}\,dx = \tfrac{1}{2}e^{x^2+9} + c.$$

Example. Evaluate $\int_1^4 3x^2(4x^3 + 5)\,dx$.

We can determine the indefinite integral by substitution.

$$u = 4x^3 + 5, \qquad du = 12x^2\,dx, \text{ and } dx = \frac{du}{12x^2}\cdot$$

The indefinite integral written in terms of u is

$$\frac{1}{4}\int u\,du = \frac{u^2}{8} + c.$$

The definite integral written in terms of x is

$$\int_1^4 3x^2(4x^3 + 5)\,dx = \left.\frac{(4x^3 + 5)^2}{8}\right|_1^4$$

$$= \frac{[4(64) + 5]^2 - [4(1) + 5]^2}{8} = \frac{68{,}121 - 81}{8}$$

$$= 8505.$$

This result can also be obtained by expressing the limits of integration in terms of u. Since $u = 4x^3 + 5$, the lower limit is $u = 4(1)^3 + 5 = 9$ and the upper limit is $u = 4(4)^3 + 5 = 261$. The definite integral is

$$\frac{1}{4}\int_{9}^{261} u\, du = \frac{u^2}{8}\bigg|_{4}^{261} = 8505.$$

Example. Evaluate $\displaystyle\int_0^3 \frac{xe^{x^2}}{3}\, dx.$

By substituting $u = x^2$ and $\dfrac{du}{2x} = dx$, and expressing the limits of integration as $u = 0$ and $u = 9$, we obtain

$$\frac{1}{6}\int_0^9 e^u\, du = \frac{e^u}{6}\bigg|_0^9 = \frac{e^9 - e^0}{6} = \frac{8103.1 - 1}{6} = 1350.3.$$

6.4.4 INTEGRATION BY PARTS

The formula for the derivative of the product of two functions is often useful in evaluating integrals. In differential calculus, the derivative of the product of the two functions, $f(x)$ and $g(x)$, is

$$\frac{d}{dx}[f(x)\cdot g(x)] = f(x)g'(x) + g(x)f'(x).$$

The integral of this derivative gives

$$f(x)\cdot g(x) = \int f(x)g'(x)\, dx + \int g(x)f'(x)\, dx,$$

which can be written as

$$\int f(x)g'(x)\, dx = f(x)g(x) - \int g(x)f'(x)\, dx. \tag{6.13}$$

This relationship can be used to determine integrals of certain functions. The following examples illustrate the technique of integration by parts.

Example. Determine $\int xe^{-x}\, dx$.

The method of integration by parts involves separating the function to be integrated into $f(x)$ and $g'(x)$. In this example we let $f(x) = x$ and $g'(x) = e^{-x}$. Formula (6.13) requires $f'(x)$ and $g(x)$. These functions are obtained by differentiating $f(x)$ and integrating $g'(x)$. Thus,

$$f(x) = x \quad \text{and} \quad f'(x) = 1.$$

$$g'(x) = e^{-x} \quad \text{and} \quad \int e^{-x}\, dx = -e^{-x}.$$

Substituting these terms in formula (6.13) gives

$$\int xe^{-x}\, dx = -xe^{-x} - \int -e^{-x}\, dx,$$

which can be integrated to yield

$$\int xe^{-x}\,dx = -xe^{-x} - e^{-x} + c = -e^{-x}(x + 1) + c.$$

Example. Determine $\int x^2e^{-x}\,dx$.
Let

$$f(x) = x^2 \quad \text{and} \quad g'(x) = e^{-x}.$$

The derivative of $f(x)$ and the integral of $g'(x)$ are

$$f'(x) = 2x, \qquad g(x) = -e^{-x}.$$

From formula (6.13) we obtain

$$\int x^2e^{-x}\,dx = -x^2e^{-x} - \int -2xe^{-x}\,dx$$

$$= -x^2e^{-x} + 2\int xe^{-x}\,dx.$$

The integral $\int xe^{-x}\,dx$ was determined in the preceding example. The integral is thus

$$\int x^2e^{-x}\,dx = -x^2e^{-x} - 2e^{-x}(x + 1) + c.$$

One of the difficult problems in applying the formula for integration by parts is to determine the terms to substitute for $f(x)$ and $g'(x)$ in the integral $\int f(x)g'(x)\,dx$. Although there are no specific rules for determining the appropriate substitution, a guideline that the student can follow is to select $g'(x)$ so that $g(x)$ can be determined and $f(x)$ so that $f'(x)$ can be determined. Referring to formula (6.13), the integral of $\int f(x)g'(x)\,dx$ is given by the $f(x)g(x)$ minus $\int g(x)f'(x)\,dx$. If the substitutions have been correctly made, $f(x)g(x)$ can be determined and $\int g(x)f'(x)\,dx$ can be integrated. If $\int g(x)f'(x)\,dx$ appears to be a more complicated integral than $\int f(x)g'(x)\,dx$, there is a good chance that the original substitutions for $f(x)$ and $g'(x)$ were incorrect.

Example. Determine $\int x \ln x\,dx$.
Let

$$f(x) = \ln x \quad \text{and} \quad g'(x) = x.$$

Then

$$f'(x) = \frac{1}{x} \quad \text{and} \quad g(x) = \frac{x^2}{2}.$$

$$\int x \ln x\,dx = \frac{x^2 \ln x}{2} - \int \frac{x}{2}\,dx$$

$$= \frac{x^2 \ln x}{2} - \frac{x^2}{4} + c.$$

Example. Determine $\int_1^3 x^2 \ln x\,dx$.

Let

$$f(x) = \ln x \quad \text{and} \quad g'(x) = x^2.$$

Then

$$f'(x) = \frac{1}{x} \quad \text{and} \quad g(x) = \frac{x^3}{3}.$$

$$\int x^2 \ln x \, dx = \frac{x^3 \ln x}{3} - \int \frac{x^2}{3} \, dx$$

$$= \frac{x^3 \ln x}{3} - \frac{x^3}{9} = \frac{x^3}{3}\left(\ln x - \frac{1}{3}\right).$$

Evaluating the definite integral for the limits of integration gives

$$\int_1^3 x^2 \ln x \, dx = \frac{x^3}{3}\left(\ln x - \frac{1}{3}\right)\Bigg|_1^3 = 9\left(\ln 3 - \frac{1}{3}\right) - \frac{1}{3}\left(\ln 1 - \frac{1}{3}\right)$$

$$9(1.09861 - 0.33333) - \tfrac{1}{3}(0.0 - 0.33333) = 6.88752 + 0.11111 = 6.99863.$$

6.4.5 INTEGRATION BY TABLES OF INTEGRALS

Determining the integrals of certain functions requires techniques of integration that are beyond the scope of business-oriented texts. Such integrals can most readily be determined by the use of tables of integrals. Table VI contains those integrals most commonly used in business and economic models. For a more extensive reference, the reader is referred to *Standard Mathematical Tables*, a widely used mathematical reference, which contains more than 500 integrals.

Example. Determine $\int \frac{dx}{x \ln x}$.

$$\int \frac{dx}{x \ln x} = \ln(\ln x) + c.$$

Example. Determine $\int (3 + 4x)^5 \, dx$.

$$\int (3 + 4x)^5 \, dx = \frac{1}{4(6)}(3 + 4x)^6 + c$$

$$= \frac{(3 + 4x)^6}{24} + c.$$

Example. Determine $\int x\sqrt{9 - x^2} \, dx$.

$$\int x\sqrt{9 - x^2} \, dx = -\tfrac{1}{3}\sqrt{(9 - x^2)^3} + c.$$

Example. Determine $\int \frac{dx}{16 - 4x^2}$.

$$\int \frac{dx}{16 - 4x^2} = \frac{1}{4} \int \frac{dx}{4 - x^2} = \frac{1}{4} \cdot \frac{1}{2(2)} \ln \left(\frac{2 + x}{2 - x} \right) + c.$$

Example. Determine $\int (4x^2 - 9)^{-1/2}\, dx$.

$$\int (4x^2 - 9)^{-1/2}\, dx = \int [4(x^2 - \tfrac{9}{4})]^{-1/2}\, dx = \tfrac{1}{2} \int (x^2 - \tfrac{9}{4})^{-1/2}\, dx$$
$$= \tfrac{1}{2} \ln [x + (x^2 - \tfrac{9}{4})^{1/2}] + c.$$

problems

1. Determine the integrals.

a. $\int 6\, dx$ b. $\int k\, dx$

c. $\int \frac{1}{3}\, dx$ d. $\int -2\, dx$

e. $\int (a + b)\, dx$ f. $\int k^2\, dx$

g. $\int dx$ h. $\int \frac{k^2}{k + 1}\, dx$

i. $\int x\, dx$ j. $\int x^2\, dx$

k. $\int x^{-2}\, dx$ l. $\int x^{1/2}\, dx$

m. $\int \frac{dx}{x^3}$ n. $\int x^{2k}\, dx$

o. $\int \sqrt{x}\, dx$ p. $\int \frac{k^2}{x^2}\, dx$

q. $\int_0^2 x^2\, dx$ r. $\int_1^3 \frac{1}{\sqrt{x}}\, dx$

s. $\int_0^5 dx$ t. $\int_2^6 x^{-1/2}\, dx$

u. $\int_1^{10} 3x^3\, dx$ v. $\int_0^5 x^{2.2}\, dx$

2. Determine the integrals.

a. $\int (3x + 4)\, dx$ b. $\int (x^2 + 2x + 6)\, dx$

c. $\int (2x^3 + x^2)\,dx$

d. $\int (x^{1/2} + x^{1/3})\,dx$

e. $\int (x + 1)(x + 1)\,dx$

f. $\int (x - 2)(x + 3)\,dx$

g. $\int_1^4 (x^2 - 2x + 3)\,dx$

h. $\int_2^5 (x - 3)(x + 1)\,dx$

i. $\int 3e^{3x}\,dx$

j. $\int 5^x\,dx$

k. $\int \frac{1}{2}e^{2x}\,dx$

l. $\int 2^{2x}\,dx$

m. $\int_0^{10} 3e^{-3x}\,dx$

n. $\int_0^{\infty} e^{-x}\,dx$

o. $\int \ln(x)\,dx$

p. $\int \ln(2x)\,dx$

q. $\int \log(3x)\,dx$

r. $\int \log_5(x)\,dx$

s. $\int_1^{10} \ln(x)\,dx$

t. $\int \log_{10}(2x)\,dx$

3. Determine the integrals. Use the method of substitution.

a. $\int (x - 1)^{5/2}\,dx$

b. $\int x(x^2 - 2)^2\,dx$

c. $\int x^2(4x^3 - 2)^9\,dx$

d. $\int xe^{x^2}\,dx$

e. $\int x^2e^{x^3}\,dx$

f. $\int 4xe^{-x^2-4}\,dx$

g. $\int \frac{1}{5x - 6}\,dx$

h. $\int \frac{dx}{4 - 3x}$

i. $\int_0^5 \frac{2x}{x^2 - 2}\,dx$

j. $\int_{-2}^2 xe^{x^2}\,dx$

k. $\int x\ln(x^2 - 6)\,dx$

l. $\int x\ln x^2\,dx$

m. $\int_0^1 \frac{2x}{x^2 + 1}\,dx$

n. $\int_0^2 x(x^2 + 1)^2\,dx$

4. Determine the integrals. Use the method of integration by parts.

a. $\int x^2\ln x\,dx$

b. $\int x^2e^{-x}\,dx$

c. $\int x^3 e^{2x}\, dx$

d. $\int x^{1/2} \ln x\, dx$

e. $\int (\ln x)^2\, dx$

f. $\int xe^{2x}\, dx$

g. $\int x3^x\, dx$

5. Calculate the area between the horizontal axis and the function $f(x) = x^2$ between $x = 0$ and $x = 5$ using formula (6.1) with $n = 5$ and $f(x)$ equal to the value of the function at the midpoint of each interval Δx. Recalculate using $n = 10$. Calculate the area using formula (6.4) and compare the result with that obtained using formula (6.1).
6. Calculate the value of the definite integral $\int_0^{10} x^3\, dx$ using formula (6.1) with $n = 5$. Compare this value to that obtained from formula (6.4).

7

application of integral calculus

7.1 area

One of the important applications of integral calculus is the evaluation of the area underlying a function. As shown in the preceding chapter, integral calculus enables one to determine the area bounded by the function and the horizontal axis between the limits of integration. Although the business and economics student may rarely have occasion to utilize this specific application, an understanding of the concept is quite useful. The concept of area is applied in probability along with other business and economic models. For this reason, we shall illustrate the concept of the definite integral as an area.

7.1.1 LINEAR FUNCTIONS

The definite integral $\int_a^b f(x)\,dx$ gives the area bounded by $f(x)$, the horizontal axis, $x = a$, and $x = b$. As an example, consider the area of the triangle shown in Fig. 7.1.

The formula for the area of a triangle is $A = \frac{1}{2}$[base · height]. The base of the triangle shown in the figure is a. The height of the triangle whose hypotenuse is described by the function $f(x) = x$ is also a. The area of the triangle is $A = \frac{1}{2}(a \cdot a) = \frac{1}{2}a^2$.

The same result can be obtained through integral calculus. Using formula (6.4) with $f(x) = x$ and the limits of integration $(0, a)$, we see that

$$A = \int_0^a x\,dx = \frac{x^2}{2}\bigg|_0^a = \frac{1}{2}a^2.$$

As expected, the two methods both provide the same result.

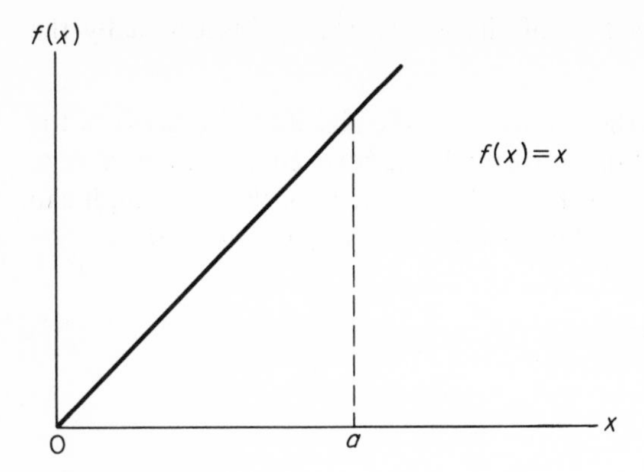

Figure 7.1

The concept of the definite integral as an area can also be illustrated by the formula for the area of a rectangle. The rectangle shown in Fig. 7.2 has height c and base $(b - a)$. The area of a rectangle is A = height · base, or $A = c(b - a)$. The function that describes the upper base of the rectangle is $f(x) = c$. The definite integral, evaluated between $x = a$ and $x = b$, gives the area bounded by $x = a$, $x = b$, the horizontal axis, and the function $f(x) = c$. This integral is $\int_a^b c\,dx = cx\Big|_a^b = c(b - a)$.

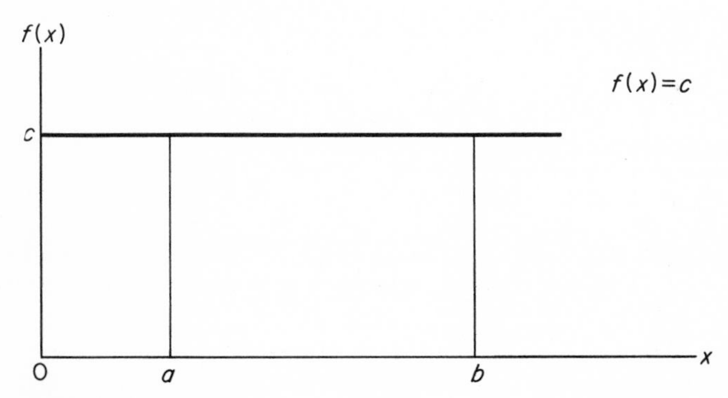

Figure 7.2

In using the definite integral for the calculation of areas, area above the horizontal axis is positive and that below the horizontal axis is negative. The process of integration automatically subtracts negative area from positive area, thus giving the net area. Consequently, if the analyst is concerned with the absolute area rather than the net area, he must determine the area beneath the horizontal axis and that above the horizontal axis separately, and then sum the absolute values of the areas. This is illustrated by the following example.

Example. Determine both the net area and the absolute area between the function $f(x) = -4 + x$ and the horizontal axis between $x = 2$ and $x = 6$. This function is shown in Fig. 7.3. From the symmetry of the diagram, it can be seen that the net area is zero. This is also shown by the integral.

$$A = \int_2^6 (-4 + x)\, dx = -4x + \frac{x^2}{2}\Big|_2^6 = -6 + 6 = 0.$$

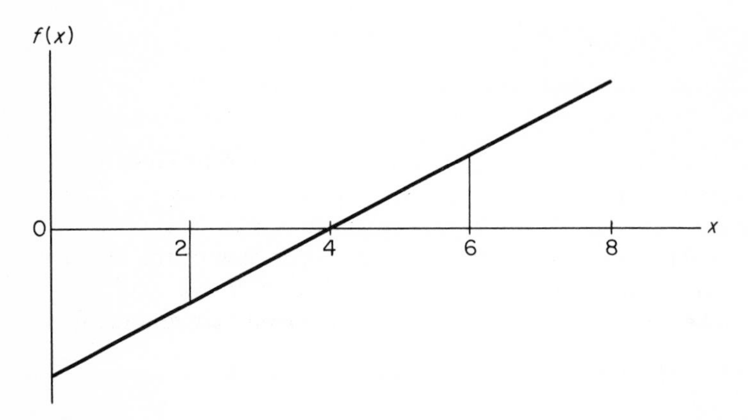

Figure 7.3

The absolute area can be found by considering the triangles separately.

$$A_1 = \int_2^4 (-4 + x)\, dx = -4x + \frac{x^2}{2}\Big|_2^4 = -2$$

$$A_2 = \int_4^6 (-4 + x)\, dx = -4x + \frac{x^2}{2}\Big|_4^6 = +2.$$

The absolute value of the area is therefore $A = |A_1| + |A_2| = 4$.

7.1.2 CURVILINEAR FUNCTIONS

The definite integral can be used to determine the area bounded by the horizontal axis and curvilinear functions. The procedure for determining the area is the same as that for linear functions, in that we again evaluate the definite integral, using the upper and lower boundaries of the area as the limits of integration. This is illustrated by the following example.

Example. Determine the area bounded by the horizontal axis, the function

$f(x) = -x^2 + 12x - 20$, $x = 2$, and $x = 10$. This function is shown in Fig. 7.4. The area is

$$A = \int_2^{10} (-x^2 + 12x - 20)\, dx = \frac{-x^3}{3} + 6x^2 - 20x \Big|_2^{10} = 85\tfrac{1}{3}.$$

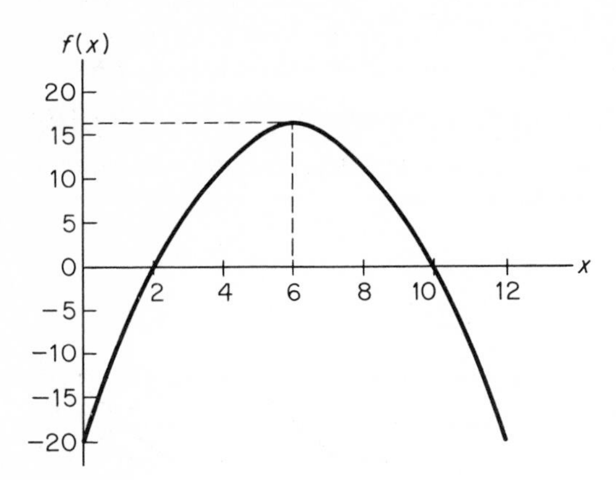

Figure 7.4

Example. Determine the net area bounded by $x = 0$ and $x = 12$ between the function $f(x) = -x^2 + 12x - 20$ and the horizontal axis. This function is plotted in Fig. 7.4.

$$A = \int_0^{12} (-x^2 + 12x - 20)\, dx = \frac{-x^3}{3} + 6x^2 - 20x \Big|_0^{12} = 48.$$

The net area between $x = 0$ and $x = 12$ is $A = 48$. Since, from the preceding example, the area between $x = 2$ and $x = 10$ is $85\frac{1}{3}$, we conclude that the negative area is $85\frac{1}{3} - 48 = 37\frac{1}{3}$. This can be verified by determining the integral between $x = 0$ and $x = 2$ and between $x = 10$ and $x = 12$.

$$A = \int_0^{2} (-x^2 + 12x - 20)\, dx + \int_{10}^{12} (-x^2 + 12x - 20)\, dx$$

$$A = \frac{-x^3}{3} + 6x^2 - 20x \Big|_0^{2} + \frac{-x^3}{3} + 6x^2 - 20x \Big|_{10}^{12}$$

$$A = -18\tfrac{2}{3} - 18\tfrac{2}{3} = -37\tfrac{1}{3}.$$

7.1.3 AREA BETWEEN FUNCTIONS

The definite integral can be used to determine the area between two functions. Given two functions $f(x)$ and $g(x)$, the area bounded by these functions and $x = a$ and $x = b$ is given by

$$A = \int_a^b [f(x) - g(x)]\,dx. \tag{7.1}$$

The integral gives the total area bounded by the two functions between $x = a$ and $x = b$ in Fig. 7.5(a). Since $f(x)$ is greater than $g(x)$ between the limits of integration, the area will be positive. This area is indicated by the shading in Fig. 7.5(a). Had we chosen to evaluate $\int_a^b [g(x) - f(x)]\,dx$, the sign prefacing the area would have been negative. The absolute value of the two areas is, however, the same.

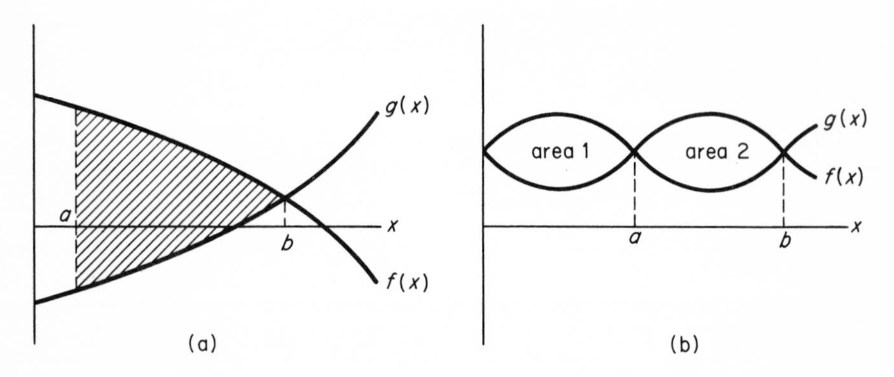

Figure 7.5

The integral gives net area; that is, area 1 − area 2, as shown in Fig. 7.5(b). Thus if absolute area is required, the integral must be evaluated as two or more integrals. In Fig. 7.5(b), the absolute value of $\int_0^a [f(x) - g(x)]\,dx$ plus the absolute value of $\int_a^b [f(x) - g(x)]\,dx$ would give the total absolute area between the functions.

Example. Determine the area between the two functions $g(x) = x - 3$ and $h(x) = 3x - x^2$. These functions are shown in Fig. 7.6.

The limits of integration are those values of x for which the two functions are equal. If we equate the two functions, we obtain

$$3x - x^2 = x - 3$$
$$x^2 - 2x - 3 = 0$$
$$x = -1, \quad x = 3.$$

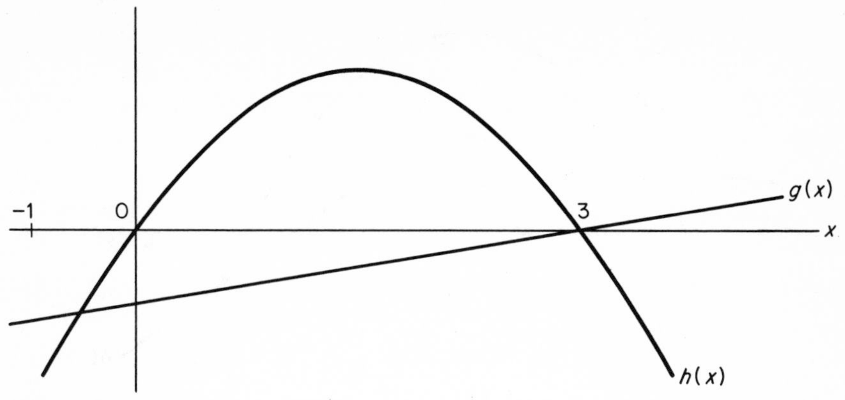

Figure 7.6

The area between the two functions for $x = -1$ to $x = 3$ is

$$A = \int_{-1}^{3} [(3x - x^2) - (x - 3)]\, dx$$

$$A = \frac{3x^2}{2} - \frac{x^3}{3} - \frac{x^2}{2} + 3x \Big|_{-1}^{3} = 10\tfrac{2}{3}.$$

We can also determine the area bounded by more than two functions. In the following example the problem is to determine the area between the quadratic function $f(x)$ and the step function $h(x)$.

Example. Determine the area bounded by $f(x)$ and $h(x)$.

$$f(x) = 4 + 5x - x^2 \qquad \text{for } 0 \leq x \leq 6$$

$$h(x) = \begin{cases} 4 + \frac{2}{3}x & \text{for } 0 \leq x \leq 3 \\ 12 - 2x & \text{for } 3 \leq x \leq 6 \end{cases}$$

These functions are shown in Fig. 7.7.

The limits of integration were determined by equating the functions and solving for $x = 0$, $x = 3.0$, and $x = 6.0$. The area is found by evaluating the definite integrals

$$A = \int_{0}^{3} [(4 + 5x - x^2) - (4 + \tfrac{2}{3}x)]\, dx + \int_{3}^{6} [(4 + 5x - x^2) - (12 - 2x)]\, dx$$

$$A = 4x + \frac{5x^2}{2} - \frac{x^3}{3} - 4x + \frac{x^2}{3}\Big|_{0}^{3} + 4x + \frac{5x^2}{2} - \frac{x^3}{3} - 12x + x^2\Big|_{3}^{6}$$

$$A = 22.5 + 6.5 = 29.0.$$

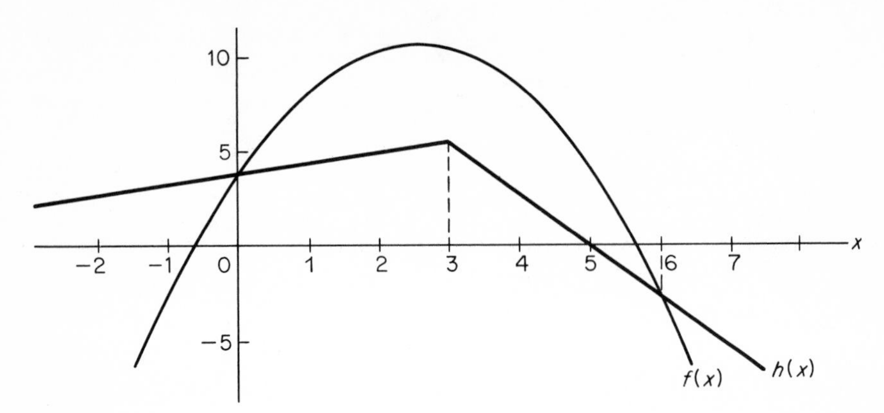

Figure 7.7

7.2 probability

One of the important uses of integral calculus in business and economics is the evaluation of probabilities of continuous random variables. Continuous random variables are defined as numerical values along a continuous line. The value of the random variable is dependent upon the results of a statistical process. If, for example, the statistical process involved weighing individuals, then the weights of individuals would be a continuous random variable. Similarly, if the statistical process was to determine the time between arrivals of customers at a service facility, then the time between arrivals of customers at the service facility would be a continuous random variable. In both of these examples the value of the random variable would depend upon the results of the statistical process. If a randomly selected individual weighed 170 pounds, the value of the random variable would be 170 pounds. Similarly, if three minutes elapsed between the arrival of customers at a service facility, then the value of the random variable would be three minutes.

The probability of a randomly selected individual's weight being between a lower and an upper limit or the probability of a customer's arriving within a certain time interval is given by a definite integral. The function that is integrated is called the *probability density function*. Figure 7.8 is a probability density function for the weights of adult males. The random variable in this figure is weight, and the density function is denoted by $f(x)$.

The probability of an event must be a value between 0 and 1, with 0 representing no chance of occurrence and 1 representing certain occurrence. If we let $f(x)$ represent a density function, then by definition

$$\int_a^b f(x)\,dx = 1, \tag{7.2}$$

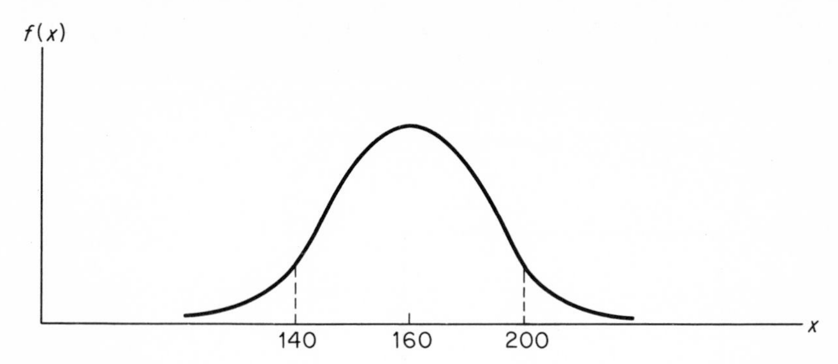

Figure 7.8

where (a, b) spans all possible values of the random variable x. Since probabilities cannot be negative, the density function must be positive.

$$f(x) \geq 0. \tag{7.3}$$

The probability that the outcome of an experiment results in the value of a random variable between a and b is given by

$$\int_a^b f(x)\,dx. \tag{7.4}$$

We can illustrate these properties by referring to Fig. 7.8. The density function in this figure describes the likelihood of occurrence of weights of adult males. Since an adult male's weight must fall someplace on the diagram, the integral of the density function over all possible weights must equal 1. The density function must be greater than or equal to zero for all weights, i.e., $f(x) \geq 0$. Finally, the probability of a randomly selected individual's weight's falling in some interval is given by the integral of the density function over that interval. In Fig. 7.8, the probability of an individual's weight's falling between 140 and 200 pounds is given by $\int_{140}^{200} f(x)\,dx$.

7.2.1 SEVERAL USEFUL PROBABILITY FUNCTIONS

We can illustrate the concept of probability as the value of a definite integral by introducing several widely used probability functions. One of these functions is the uniform density function or uniform distribution. The uniform distribution describes an experiment in which the value of the random variable assumes some value in the domain $a \leq x \leq b$. All values of x in the domain are equally likely, and the density function is

$$f(x) = \frac{1}{b-a} \qquad \text{for } a \leq x \leq b. \tag{7.5}$$

The density function has the value of zero for values outside the domain. This distribution is shown in Fig. 7.9.

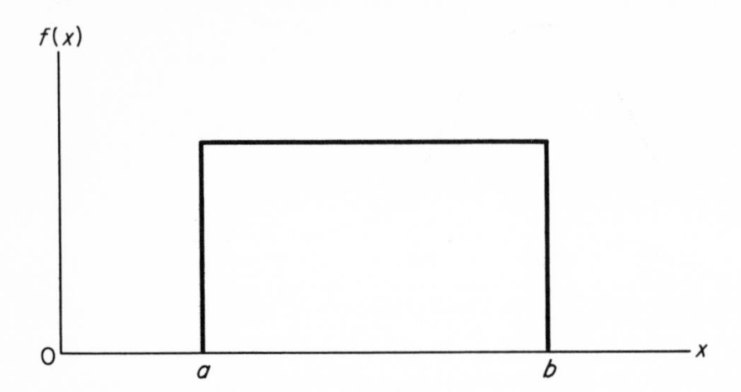

Figure 7.9

The density function is $f(x) = \dfrac{1}{b - a}$ for $a \leq x \leq b$. The definite integral of $f(x)$ with limits of integration a and b is equal to 1, as shown below.

$$\int_a^b \frac{1}{b - a}\, dx = \left.\frac{x}{b - a}\right|_a^b = \frac{b - a}{b - a} = 1.$$

The use of this function is illustrated by the following examples.

Example. Busses on a certain route run every half hour. What is the probability that a man arriving at a random time to catch a bus will have to wait at least 20 minutes?

The random variable t = time to wait until the next bus is uniformly distributed for $0 \leq t \leq 30$. The probability that the man must wait at least 20 minutes is

$$P(t > 20) = \int_{20}^{30} \frac{1}{30}\, dt = \left.\frac{t}{30}\right|_{20}^{30} = \frac{1}{3}.$$

Example. A four-digit random number table contains a selection of random numbers varying from 0000 to 9999. If these numbers are described by a uniform distribution, determine the probability of selecting a random number whose value is between 400 and 500, inclusive.

The probability density function for the random numbers is

$$f(x) = \frac{1}{9999.5 - (-0.5)} = \frac{1}{10{,}000} \qquad \text{for } 0 \leq x \leq 9999.$$

This function is derived from the fact that we represent a random number, $x = a$, as the midpoint of the interval $x = a \pm 0.5$. The probability of a random number's falling in the interval from 400 to 500, inclusive, is thus

$$P(400 \leq x \leq 500) = \int_{399.5}^{500.5} \frac{1}{10{,}000}\,dx = \left.\frac{x}{10{,}000}\right|_{399.5}^{500.5} = \frac{101}{10{,}000}.$$

Another useful probability model is the exponential density function. One use of the exponential density function is in describing the arrival of an individual or item at a service facility. For example, the time between arrivals of customers at a service station is often described by the exponential probability distribution.

The probability density function for the exponential distribution is

$$f(t) = ke^{-kt} \qquad \text{for } 0 \leq t \leq \infty. \tag{7.6}$$

In the density function, t represents time between arrivals, $e = 2.71828$ is a constant, and k is the average number of arrivals per period. The exponential distribution is shown in Fig. 7.10.

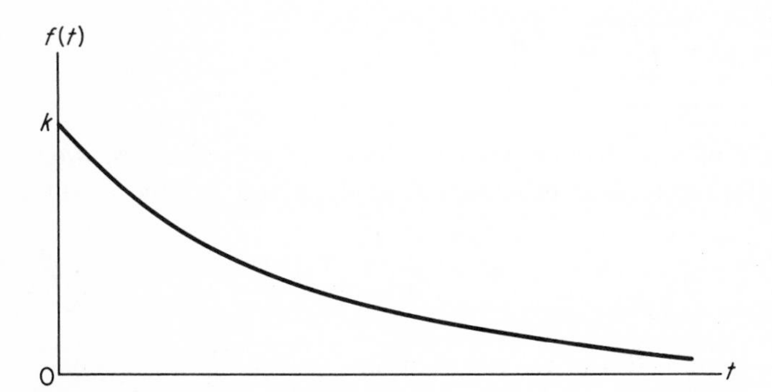

Figure 7.10

The probability of an arrival within a time interval $t \leq T$ is $P(t \leq T) = \int_0^T ke^{-kt}\,dt$. The exponential density function satisfies the requirements for a density function, since $f(t) \geq 0$, and

$$\int_0^\infty ke^{-kt}\,dt = \left.-e^{-kt}\right|_0^\infty = 0 - (-1) = 1.$$

This distribution is illustrated by the following examples.

Example. Customers are known to arrive at a service station according to the exponential distribution with an average of $k = 10$ arrivals per hour.

Determine the probability of an arrival within $T = 0.2$ hour.

$$P(t \le 0.2) = \int_0^{0.2} 10e^{-10t}\,dt = -e^{-10t}\Big|_0^{0.2} = -e^{-2} + 1$$
$$= -0.135 + 1 = 0.865.$$

The probability of an arrival within 0.2 hour is 0.865.

Example. Aircraft have been observed to arrive at a certain airport according to the exponential distribution with $k = 20$ arrivals per hour. Determine the probability of an arrival of an aircraft within a three-minute period (i.e., $\frac{1}{20}$ of an hour).

$$P(t \le \tfrac{1}{20}\text{ hr}) = \int_0^{1/20} 20e^{-20t}\,dt$$
$$= -e^{-20t}\Big|_0^{1/20} = -e^{-1} + 1 = -0.368 + 1$$
$$= 0.632.$$

Example. The time between breakdowns of a certain part is described by the exponential distribution with $k = 2$ breakdowns per year. Determine the probability of a breakdown in a nine-month period.

$$P(t \le \tfrac{3}{4}) = \int_0^{3/4} 2e^{-2t}\,dt = -e^{-2t}\Big|_0^{3/4}$$
$$= -e^{-1.5} + 1 = -0.223 + 1 = 0.777.$$

A third example of a probability distribution is the *normal distribution.* This distribution is widely used in statistical analysis and has the density function

$$f(x) = \frac{1}{\sigma\sqrt{2\pi}}\, e^{-\frac{1}{2}[(x-u)/\sigma]^2} \qquad \text{for } -\infty < x < \infty, \tag{7.7}$$

where σ and u are parameters of the distribution representing the standard deviation and mean, respectively. This distribution has the familiar bell shape, as is shown by Fig. 7.11.

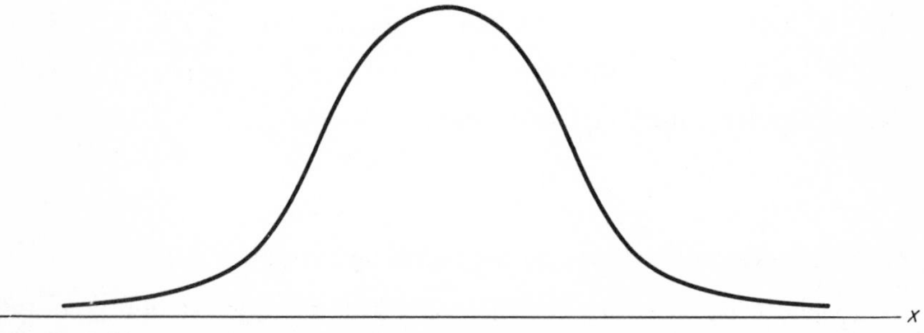

Figure 7.11

The probability of the random variable's falling in the interval $a \leq x \leq b$ is given by

$$P(a \leq x \leq b) = \int_a^b \frac{1}{\sigma\sqrt{2\pi}} e^{-\frac{1}{2}[(x-u)/\sigma]^2} dx.$$

This integral cannot be evaluated by elementary methods. Because of the normal distribution's widespread applicability in statistical analysis, tables that give the value of the definite integral are widely available. Those individuals who study statistics will become familiar with these tables.

The uniform, exponential, and normal are widely used continuous distributions. Distributions such as the chi square, the t, the F, and the gamma are also important in statistical theory and analysis. Discussions of these probability distributions are available in advanced statistical textbooks.

7.2.2 MEAN AND VARIANCE OF A PROBABILITY DENSITY FUNCTION

The two measures of a probability density function that are widely used are the mean or expected value and the variance. The mean or expected value of a random variable x is denoted by the symbol $E(x)$ or u. The expected value of the random variable x is defined as

$$E(x) = \int_{\text{all } x} x f(x)\, dx, \tag{7.8}$$

where $f(x)$ is again the density function and x is the random variable.

The expected value of the random variable is interpreted as the average value of the random variable. Thus, if the random variable x describes time between arrivals, the expected value of the random variable is the average time between arrivals. Similarly, if the random variable x describes the weights of individuals, the expected value of the random variable is the average weight of the population of individuals. The use of (7.8) is illustrated by the following examples.

Example. An individual arrives according to the uniform distribution to board a bus, which arrives every 30 minutes. Determine his expected waiting time.

The probability density function describing the arrival of the individual is $f(t) = \frac{1}{30}$. The expected value is

$$E(t) = \int_{\text{all } t} t f(t)\, dt$$

$$E(t) = \int_0^{30} t \frac{1}{30}\, dt = \frac{t^2}{60}\Big|_0^{30} = \frac{900}{60} = 15 \text{ minutes.}$$

Example. Derive the formula for the expected value of the uniform distribution.

$$E(x) = \int_{\text{all } x} x\, f(x) dx$$

$$E(x) = \int_a^b x \left(\frac{1}{b-a}\right) dx = \left.\frac{x^2}{2(b-a)}\right|_a^b = \frac{b^2 - a^2}{2(b-a)}$$

$$E(x) = \frac{(b-a)(b+a)}{2(b-a)} = \frac{1}{2}(b+a). \tag{7.9}$$

Example. If aircraft arrive at an airport according to the exponential distribution with $k = 20$ arrivals per hour, determine the average time between arrivals.

The probability density function that describes the arrivals is $f(t) = 20e^{-20t}$. The expected value of the distribution is

$$E(t) = \int_0^\infty t(20e^{-20t})\, dt = -20te^{-20t} - \left.\frac{e^{-20t}}{20}\right|_0^\infty = e^{-20t}(-20t - \tfrac{1}{20})\Big|_0^\infty$$

$$E(t) = \tfrac{1}{20} \text{ hour.}$$

Example. Derive the formula for the expected value of the exponential distribution.

$$E(t) = \int_0^\infty kte^{-kt}\, dt = \left| -kte^{-kt} - \frac{e^{-kt}}{k}\right|_0^\infty$$

$$E(t) = \underset{t\to\infty}{\text{limit}} \left|\frac{1}{e^{kt}}\left[-kt - \frac{1}{k}\right]\right|_0^\infty = \frac{1}{k}\cdot \tag{7.10}$$

The variance of a random variable provides a measure of the magnitude of the variation of the values of the random variable about the expected value of the random variable. Figure 7.12 illustrates the concept of variance for the uniform distribution. The distributions in both Figs. 7.12(a) and 7.12(b) are uniform, and both have identical expected values. The values of the random variable in Fig. 7.12(a) are more closely distributed about the mean in Fig. 7.12(a) than are those of Fig. 7.12(b). Since the variance provides a measure of the magnitude of the variation of the values of the random variable about the mean, the variance of the distribution in Fig. 7.12(a) is less than that of the distribution in Fig. 7.12(b). The formula for the variance of a continuous random variable is

$$\text{Var}(x) = E(x^2) - [E(x)]^2. \tag{7.11}$$

The formula indicates that the variance is calculated in two steps. The definite integral $E(x^2) = \int_{\text{all } x} x^2 f(x)\, dx$ is first determined. The square of the expected value, $E(x)^2$, is then subtracted from the definite integral.

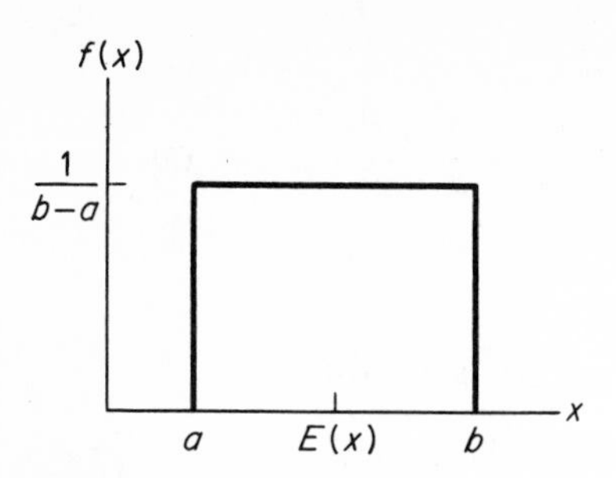

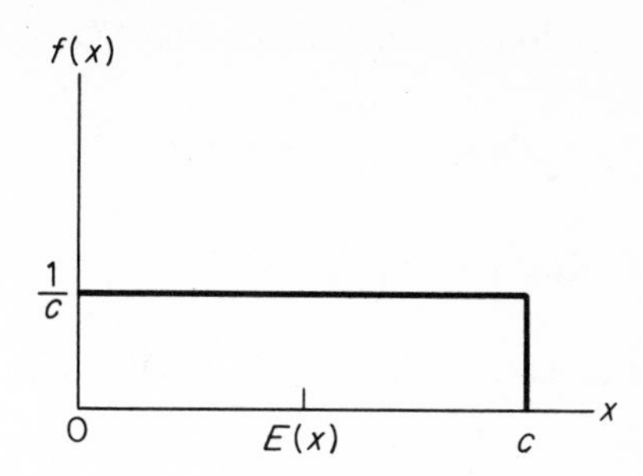

Figure 7.12

Example. Determine the variance of the uniform distribution of the example on p. 200 concerning the arrival of an individual at a bus station.

The density function for this example was $f(t) = \frac{1}{30}$. The variance is calculated in two steps. We first determine

$$E(x^2) = \int_0^{30} \frac{t^2}{30}\,dt = \frac{t^3}{90}\bigg|_0^{30} = 300.$$

Next, we square the expected value of x, $[E(x)]^2$.

$$[E(x)]^2 = (15)^2 = 225.$$

The difference between the two quantities is the variance.

$$\text{Var}(x) = 300 - 225 = 75.$$

Example. Derive the formula for the variance of the uniform distribution.

The density function for the uniform distribution is $f(x) = \dfrac{1}{b-a}$, and the expected value of the distribution is $E(x) = \dfrac{b+a}{2}$.

$$E(x^2) = \int_a^b \frac{x^2}{b-a}\,dx = \frac{x^3}{3(b-a)}\bigg|_a^b = \frac{b^3 - a^3}{3(b-a)},$$

and the variance is found by subtracting the square of the mean from $E(x^2)$. Thus,

$$\text{Var}(x) = \frac{b^3 - a^3}{3(b-a)} - \frac{(b+a)^2}{4}. \tag{7.12}$$

The variance of the exponential distribution is calculated by using formula (7.11). On the basis of this formula and applying the integral from the table of integrals, we can derive the formula for the variance of the exponential distribution. We first determine $E(x^2)$.

$$E(x^2) = k \int_0^\infty x^2 e^{-kx}\, dx = k \left| \frac{-x^2 e^{-kx}}{k} - \frac{2}{k} \cdot \frac{e^{-kx}}{k^2}(kx - 1) \right|_0^\infty$$

$$E(x^2) = k\left(\frac{2}{k^3}\right) = \frac{2}{k^2}.$$

The expected value of x is $E(x) = 1/k$. The variance is thus

$$\text{Var}(x) = \frac{2}{k^2} - \frac{1}{k^2} = \frac{1}{k^2}. \tag{7.13}$$

Example. Aircraft arrive at an airport according to the exponential distribution with $k = 20$ arrivals per hour. Determine the expected time between arrivals of successive aircraft and the variance of the time between arrivals.

This example was considered earlier. The formula $E(t) = 1/k$ gives the expected time between arrivals. Thus,

$$E(t) = \frac{1}{k} = \frac{1}{20} = 0.05.$$

The variance of the time between arrivals can be calculated by the formula $\text{Var}(t) = 1/k^2$. Thus,

$$\text{Var}(t) = \frac{1}{k^2} = \frac{1}{400} = 0.0025.$$

7.2.3 TAILOR-MADE PROBABILITY DISTRIBUTIONS

Conventional probability distributions such as the uniform, exponential, and normal distributions often do not accurately model a real-world phenomenon. When the more common probability distributions do not apply to a particular problem, it is necessary to develop a probability distribution that is tailor-made for the problem. The requirements of a density function, it will be remembered, are that the density function be greater than or equal to zero for all values of the random variable and that the definite integral of the density function over all values of the random variable is one. If these requirements are satisfied and if the density function is selected such that it describes the probability of occurrence of values of the random variable, then the tailor-made distribution has been correctly determined. The use of integral calculus in determining the parameters of a tailor-made distribution is illustrated by the following examples.

Example. A random variable is known to have the distribution shown below. Determine the value of k in the density function $f(x) = kx^2$ and the mean and variance of the distribution.

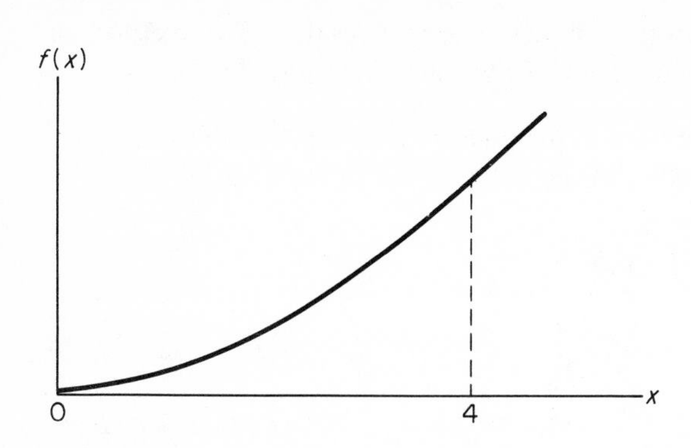

By definition, the integral of a density function is 1. Thus,

$$\int_0^4 kx^2\,dx = 1$$

$$k\left|\frac{x^3}{3}\right|_0^4 = 1$$

$$k = \tfrac{3}{64}.$$

The mean or expected value of the distribution is given by

$$E(x) = \int_0^4 x\left[\frac{3}{64}x^2\right]dx = \int_0^4 \frac{3}{64}x^3\,dx = \frac{3}{64}\left|\frac{x^4}{4}\right|_0^4$$

$$= 3.$$

The variance is found by subtracting $[E(x)]^2$ from $E(x^2)$.

$$E(x^2) = \int_0^4 x^2\left[\frac{3}{64}x^2\right]dx = \int_0^4 \frac{3}{64}x^4\,dx = \frac{3}{64}\left|\frac{x^5}{5}\right|_0^4$$

$$E(x^2) = 9.6$$

$$\text{Var}(x) = E(x^2) - [E(x)]^2 = 9.6 - 9.0 = 0.6.$$

7.2.4 MEDIAN OF A PROBABILITY DISTRIBUTION

In addition to the expected value and variance of a random variable, the median of a probability distribution is of importance in statistical analysis. The median of a probability distribution is the value of the random variable m for which the probability is 0.5 that $x \leq m$ and 0.5 that $x \geq m$. Written in terms of the definite integral, the median is defined as the value of x such that

$$\int_a^m f(x)\,dx = \tfrac{1}{2}, \qquad (7.14)$$

where $a \leq x \leq b$ is the domain of the random variable. The method of determining the median is illustrated by the following example.

Example. The density function in the preceding example was $f(x) = \frac{3}{64}x^2$ for $0 \leq x \leq 4$. Determine the median value of the random variable.

$$\int_0^m \tfrac{3}{64}x^2\,dx = \tfrac{1}{2}.$$

$$\left.\frac{x^3}{64}\right|_0^m = \frac{1}{2}$$

$$m^3 = 32$$

$$m = 3.178.$$

Example. A probability density function is described by

$$f(x) = \frac{x}{50} \qquad \text{for } 0 \leq x \leq 10.$$

Determine the expected value, variance, and median of the distribution.

$$E(x) = \int_0^{10} \frac{x^2}{50}\,dx = \left.\frac{x^3}{150}\right|_0^{10} = 6.67$$

$$E(x^2) = \int_0^{10} \frac{x^3}{50}\,dx = \left.\frac{x^4}{200}\right|_0^{10} = 50$$

$$\text{Var}(x) = E(x^2) - [E(x)]^2 = 50.0 - 44.6 = 5.4.$$

Median:

$$\int_0^m \frac{x}{50}\,dx = \frac{1}{2}$$

$$\left.\frac{x^2}{100}\right|_0^m = \frac{1}{2}$$

$$m^2 = 50$$

$$m = 7.07.$$

7.3 economic applications

The relationship between the integral and the derivative forms the basis for numerous economic models. We shall illustrate basic models in cost theory, capital formulation, and mathematics of finance which utilize this relationship.

7.3.1 MARGINAL AND TOTAL COST FUNCTIONS

We have shown that the marginal cost function is given by the derivative of the total cost function. This relationship is reversible; i.e., the integral of the marginal cost function is the total cost function.

$$TC = \int MC = TVC + FC. \tag{7.15}$$

The integral of marginal cost gives total variable cost plus the constant. The constant represents fixed cost, which does not vary with changes in the quantity of goods produced. The relationship between total cost and marginal cost is illustrated by the following examples.

Example. The cost of producing a simple component is composed of a fixed cost and a variable cost. If the fixed cost is k and the variable cost is c per unit, show that the total cost is given by $TC = cq + k$.

The cost of producing an additional unit is c. The marginal cost function is, therefore, $MC = c$. The integral of this function is

$$TC = \int c\, dq = cq + k.$$

Total cost is composed of total variable cost plus fixed cost.

Example. The cost of manufacturing an item consists of \$1000 fixed cost and \$1.50 per unit variable cost. Determine the total cost function and the cost of manufacturing 2000 units.

$$TC = \int 1.50\, dq = 1.50q + 1000.$$

The cost of manufacturing 2000 units is

$$TC = 1.50(2000) + 1000 = \$4000.$$

Example. Assume that the marginal cost of manufacturing a product is the following function of output:

$$MC = 3q^2 - 4q + 5.$$

Determine the total cost function.

$$TC = \int (3q^2 - 4q + 5)\, dq = q^3 - 2q^2 + 5q + k,$$

where k is the fixed cost of manufacturing the product.

The definite integral can be used to determine the total variable cost of producing some quantity of output. The total variable cost of producing the ath to the bth units is given by $\int_{(a-1)}^{b} MC$. This is illustrated by the following examples.

Example. The marginal cost function for a certain commodity is $MC = 3q^2 - 4q + 5$. Determine the cost of producing the eleventh through the fifteenth units, inclusive.

$$\begin{aligned} TVC &= \int_{10}^{15} (3q^2 - 4q + 5)\, dq = \left| q^3 - 2q^2 + 5q \right|_{10}^{15} \\ &= (15)^3 - 2(15)^2 + 5(15) - (10)^3 + 2(10)^2 - 5(10) = \$2150. \end{aligned}$$

The evaluation of the definite integral gives the same result that would be obtained by evaluating the total cost function for b units and the total cost function for $(a - 1)$ units and subtracting the two costs. This difference is the incremental cost of producing the ath to the bth units of output. For the above example,

$$\begin{aligned} TC(15 \text{ units}) &= (15)^3 - 2(15)^2 + 5(15) + k \\ TC(10 \text{ units}) &= (10)^3 - 2(10)^2 + 5(10) + k \\ TC(15) - TC(10) &= \$2150. \end{aligned}$$

The cost of producing units 11 through 15 is \$2150.

7.3.2 CAPITAL FORMATION AND CONTINUOUS FINANCIAL PROCESSES

The growth rate for a continuous function was discussed in Chapter 5. The formula for the rate of growth was

$$r = \frac{f'(t)}{f(t)}. \tag{5.44}$$

If the rate of growth is known rather than the function, we can determine the function by integrating formula (5.44). To demonstrate the methodology, assume that \$1 is invested at r percent and the interest is compounded continuously for a period of t years. Determine the function that describes the continuous growth.

We rewrite (5.44) to obtain

$$f'(t) = rf(t). \tag{7.16}$$

Since r is a constant, the integral of the function is

$$\int f'(t)\, dt = r \int f(t)\, dt. \tag{7.17}$$

From (7.16) we observe that the derivative of the function $f'(t)$ is equal to r times the function $f(t)$. The only function that has the property that the derivative of the function is equal to a constant times the function is the exponential function. We thus assume that $f(t) = e^{rt}$. Integrating both sides of (7.17) gives

$$\int f'(t)\, dt = r \int e^{rt}\, dt,$$

or

$$f(t) = e^{rt}. \tag{7.18}$$

We can verify that $f(t) = e^{rt}$ by differentiating $f(t)$.

$$f'(t) = re^{rt},$$

which can be expressed as

$$f'(t) = rf(t),$$

since $f(t) = e^{rt}$. The function that describes the amount of the investment at the end of period t is thus

$$A = Pe^{rt}, \tag{7.19}$$

where P is the value of the function when $t = 0$, i.e., the initial investment. This formula is derived in Chapter 5, based upon the concept of the limit and is given by (5.34) as $A = Pe^{it}$. The use of this formula in models of capital formation is illustrated by the following examples.

Example. An investment of $100 at 10 percent is compounded continuously for five years. Determine the amount of the investment at the end of the fifth year.

$$A = \$100e^{0.10t} = \$100e^{0.5} = \$164.90.$$

Example. Determine the amount which must be invested at 8 percent to accumulate to $500 at the end of ten years.

$$P = \$Ae^{-0.80t} = \$500e^{-0.08(10)} = \$224.$$

Formula (7.19) is used to determine the amount A to which an investment of P will accumulate when invested at interest rate r and compounded continuously for t years. An interesting extension of (7.19) occurs when we determine the integral of the function. This integral gives the amount to which a series of n equal payments of p invested at interest r and compounded continuously will accumulate. The reader will recognize this as the amount of an annuity assuming continuous compounding. The integral of (7.19) is

$$A_n = \int_0^t pe^{rx}\,dx,$$

where x is substituted for t as the variable of integration. Performing the integration gives

$$A_n = \left.\frac{pe^{rx}}{r}\right|_0^t$$

$$A_n = \frac{p(e^{rt} - 1)}{r}. \tag{7.20}$$

This formula is similar to (5.38) in which periodic compounding rather than continuous compounding is assumed. The only difference between the discrete and the continuous case is that $(1 + i)^t$ is replaced by e^{rt}. Applications of the formula are provided by the following examples.

Example. The president of a certain country has called for sacrifices in the consumption of consumer goods in order to shift production to capital goods. The ten-year plan is to increase the capital equipment of the country at a rate of 10 percent per year. The present stock of capital equipment is valued at $10 billion. Assuming continuous compounding, determine the amount that must be invested during the ten-year period.

The total investment during the ten years can be determined from (7.20)

$$A_n = \frac{\$10(e^{0.10(10)} - 1)}{0.10} = \frac{10(2.718 - 1)}{0.10}$$

$$A_n = \$171.8 \text{ billion.}$$

Example. For the preceding example, determine the investment required during the ninth and tenth years.

The total investment during the final two years can be determined from (7.19).

$$A_n = \int_8^{10} 10e^{0.10t}\,dt = \left.\frac{10e^{0.10t}}{0.10}\right|_8^{10}$$

$$A_n = 100(e^{1.0} - e^{0.8}) = 100(2.718 - 2.226)$$

$$A_n = \$49.2 \text{ billion.}$$

Example. An individual deposits $1000 at the end of each year in a savings account. The account pays 6 percent interest compounded daily. Determine the amount on deposit at the end of 10 years.

In working this problem, we shall assume continuous compounding rather than 365 compounding periods per year. Based upon this approximation, the amount is

$$A_n = \frac{P}{r}(e^{rt} - 1)$$

$$A_n = \frac{1000}{0.06}(e^{0.6} - 1) = \frac{1000}{0.06}(1.822 - 1)$$

$$A_n = \$1{,}370.$$

Example. In the preceding example, calculate the amount, assuming that the $1000 was deposited at the first of the year rather than the last of the year. We can determine the amount by multiplying the formula by e^r.

$$A_n = \frac{P}{r}(e^{rt} - 1)(e^r)$$

$$A_n = 1370(e^{0.06}) = 1370(1.0618)$$
$$A_n = \$1464.$$

The formula for the amount of an annuity if periodic compounding is assumed can be modified to account for payment at the first of the period rather than the last by multiplying the formula by $(1 + i)$. Thus, the formula for periodic compounding is

$$A_n = \frac{P}{r}[(1 + i)^t - 1)](1 + i), \tag{7.21}$$

and the formula for the amount of an annuity with the payment at the first of the period if continuous compounding is assumed is

$$A_n = \frac{P}{r}[e^{rt} - 1]e^{rt}. \tag{7.22}$$

Integral calculus can be used to derive the formula for the present value of an annuity. Based upon continuous compounding, the present value of an amount A received t years hence is

$$P = Ae^{-rt}. \tag{5.35}$$

Instead of receiving a single payment of A at the end of t years, consider the case in which we receive a payment of A at the end of each year. The present value of this series of payments is found by integrating (5.35) between the limits of integration of $t = 0$ and $t = t$. Evaluation of this definite integral gives the present value of an annuity if continuous compounding is assumed. The formula is

$$PV = \int_0^t Ae^{-rx}\,dx$$

$$PV = \frac{A}{r}[-e^{-rt}\Big|_0^t$$

$$PV = \frac{A}{r}[-e^{-rt} - (-e^0)]$$

$$PV = A\,\frac{(1 - e^{-rt})}{r}. \tag{7.23}$$

This formula is similar to the formula for the present value of an annuity if a discrete number of compounding periods is assumed,

$$PV = p\left[\frac{1 - (1 + i)^{-n}}{i}\right]. \tag{5.42}$$

Example. A research and development firm has patented a new photocopy process, which they have licensed to a major manufacturer. The agreement calls for the annual payment of \$50,000 for 10 years. If the discount rate is

8 percent and continuous compounding is assumed, determine the present value of this agreement.

$$PV = 50{,}000\,\frac{(1 - e^{-0.08(10)})}{0.08}$$

$$PV = \frac{50{,}000(1 - 0.4493)}{0.08}$$

$$PV = \$344{,}200.$$

problems

1. Determine the area bounded by the horizontal axis and the following functions for the specified domain of x.
   a. $f(x) = 2x$ for $0 \leq x \leq 6$
   b. $f(x) = 6 - x$ for $0 \leq x \leq 4$
   c. $f(x) = 3 + 2x$ for $-1 \leq x \leq 10$
   d. $f(x) = 10 - x$ for $-10 \leq x \leq 0$
   e. $f(x) = 3x + 4$ for $0 \leq x \leq 8$
   f. $f(x) = -6 + x$ for $0 \leq x \leq 6$
   g. $f(x) = 0.5 + 0.5x$ for $5 \leq x \leq 10$
   h. $f(x) = 15 - 15x$ for $0 \leq x \leq 1$
   i. $f(x) = 15 - 15x$ for $1 \leq x \leq 2$
   j. $f(x) = -60 + 10x$ for $0 \leq x \leq 10$
   k. $f(x) = 6$ for $5 \leq x \leq 10$
   l. $f(x) = x^2$ for $0 \leq x \leq 5$
   m. $f(x) = -x^2$ for $0 \leq x \leq 5$
   n. $f(x) = 2 + 3x + x^2$ for $0 \leq x \leq 6$
   o. $f(x) = 3 + 6x - x^2$ for $0 \leq x \leq 6$
   p. $f(x) = 100 - 15x + x^2$ for $5 \leq x \leq 10$
   q. $f(x) = 1 + x + x^2 + x^3$ for $0 \leq x \leq 10$
   r. $f(x) = 6 - x + x^2 - x^3$ for $0 \leq x \leq 5$
   s. $f(x) = 3 + x + x^2 - x^3$ for $0 \leq x \leq 5$
   t. $f(x) = 0.5e^{-0.5x}$ for $0 \leq x \leq 20$
   u. $f(x) = 4e^{6x}$ for $0 \leq x \leq 10$
   v. $f(x) = 6e^{-6x}$ for $0 \leq x \leq 1000$
   w. $f(x) = 10e^{-10x}$ for $0 \leq x \leq 0.50$
   x. $f(x) = 0.5e^{-0.5x}$ for $1 \leq x \leq 2$
   y. $f(x) = 0.3e^{-0.3x}$ for $0.5 \leq x \leq 1.5$
   z. $f(x) = 0.1e^{-0.1x}$ for $0.1 \leq x \leq 0.2$
2. Determine the area between $f(x)$ and $g(x)$ for the specified domain of x.
   a. $f(x) = 3 + 2x$, $g(x) = x$ for $0 \leq x \leq 5$

b. $f(x) = 6, g(x) = 3$ for $0 \leq x \leq 10$
c. $f(x) = 10 - x, g(x) = 2 + x$ for $0 \leq x \leq 4$
d. $f(x) = 25 - 2x, g(x) = -25 + 3x$ for $0 \leq x \leq 5$
e. $f(x) = 5 + x, g(x) = 5 - x$ for $0 \leq x \leq 10$
f. $f(x) = 6 + 5x - x^2, g(x) = 6 - 5x + x^2$ for $0 \leq x \leq 10$
g. $f(x) = 10 - 2x + x^2, g(x) = 10 + 2x - x^2$ for $0 \leq x \leq 4$
h. $f(x) = 15 - 3x + x^2, g(x) = 5 - 3x + x^2$ for $0 \leq x \leq 6$
i. $f(x) = 6 + x + x^2, g(x) = 5 + x + x^2$ for $0 \leq x \leq 10$
j. $f(x) = 15 - 2x + x^2, g(x) = -15 + 2x - x^2$ for $5 \leq x \leq 10$
k. $f(x) = 20 - 2x + x^2, g(x) = 10$ for $2 \leq x \leq 6$
l. $f(x) = 10 + 4x - x^2, g(x) = 4$ for $3 \leq x \leq 6$
m. $f(x) = 20 - x + x^2, g(x) = -20 + x - x^2$ for $-4 \leq x \leq 10$

3. Determine the value of k such that the following functions are probability density functions.
   a. $f(x) = k(6), 0 \leq x \leq 10$
   b. $f(x) = k(3), 0 \leq x \leq 5$
   c. $f(x) = k(3 - x), 0 \leq x \leq 3$
   d. $f(x) = k(2 + 3x), 0 \leq x \leq 10$
   e. $f(x) = k(-3 + x), 3 \leq x \leq 10$
   f. $f(x) = k(10 - 2x), 2 \leq x \leq 5$
   g. $f(x) = k(x^2 - x), 1 \leq x \leq 4$
   h. $f(x) = kx^2, 0 \leq x \leq 5$
   i. $f(x) = k(3x + x^2), 0 \leq x \leq 6$
   j. $f(x) = k(10 - 2x + x^2), 10 \leq x \leq 15$
   k. $f(x) = ke^{-3x}, 0 \leq x \leq \infty$
   l. $f(x) = k(x^2 + 2x + 10), 0 \leq x \leq 1$
   m. $f(x) = k(5x + 5x^2), 1 \leq x \leq 3$
4. Determine the mean or expected value for each of the functions in Question 3.
5. Determine the variance for each of the functions in Question 3.
6. Determine the median for each of the functions in Question 3.
7. For each of the following problems develop a probability density function. Determine the mean and variance of the probability distribution and determine the probability of the event specified in the problem.
   a. An automated machine completes an operation every two minutes. What is the probability that an inspector randomly arriving will have to wait 30 seconds or less for the part?
   b. The probability of failure of a certain part is described by the density function $f(t) = kt^2$, for $0 \leq t \leq 10$. Determine the probability of failure for $t \leq 5$.
   c. The time between arrivals of ships at a west coast port is given

by the exponential probability density function with the mean time between arrival equal to 0.5 day. Determine the probability of no arrivals in 0.25 day.

d. The time between telephone calls at a resort hotel is distributed according to the exponential distribution with an expected time between calls of four minutes. Determine the probability of a call within two minutes.
e. The life of a certain part is distributed according to the uniform distribution with $10 \leq t \leq 20$. Determine the probability of failure for $15 \leq t \leq 18$.
f. The probability of failure of a radio tube is described by the probability density function $f(t) = kt^2$ for $0 \leq t \leq 120$ months. Determine the probability of the tubes failing within 60 months.

appendix A

establishing function

There are several methods in which functions can be established such that the function passes through data points. The most straightforward and easily understandable method of establishing these functions is through the solution of simultaneous equations.

A.1 linear functions

As a first illustration, consider the problem of establishing a linear function. Two data points are required to define a linear function. We shall consider the data points ($x = 1, y = 4$) and ($x = 4, y = 2$). It is customary to enclose the data points in parentheses with the independent or x variable preceding the dependent or y variable and the two variables separated by commas. Thus the two data points are customarily written as (1, 4) and (4, 2). The general form of the linear function is $y = a + bx$. To determine the values of a and b we substitute the two data points for x and y and solve the resulting two equations simultaneously. Thus,

$$4 = a + b(1)$$
$$2 = a + b(4)$$

must be solved simultaneously for a and b.

Subtracting the second equation from the first gives

$$2 = -3b$$

and $b = -\frac{2}{3}$. Substituting for b in either equation and solving for a gives $a = 4\frac{2}{3}$. This function is shown in Fig. A.1. A second sample follows.

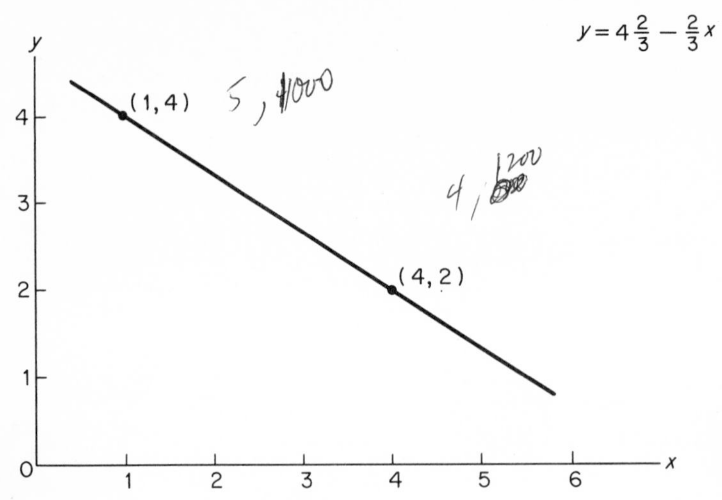

Figure A.1. Linear Function

Example. Determine the linear function that passes through the data points (2, −3) and (10, 17). The two equations are

$$-3 = a + b(2)$$
$$17 = a + b(10),$$

which when solved simultaneously give

$$y = -8 + 2\tfrac{1}{2}x.$$

A.2 quadratic functions

A quadratic, or second degree, function is of the form $y = a + bx + cx^2$. Three data points are necessary to define a quadratic function. The three data points are substituted into the general form of the quadratic function, and the resulting three equations are solved simultaneously for the parameters a, b, and c. The procedure is illustrated by the following example.

Example. Determine the quadratic function that passes through the data points (1, 6), (3, 4), and (8, 7). These data points are shown in Fig. A.2. The three equations are

$$6 = a + b(1) + c(1)^2$$
$$4 = a + b(3) + c(3)^2$$
$$7 = a + b(8) + c(8)^2,$$

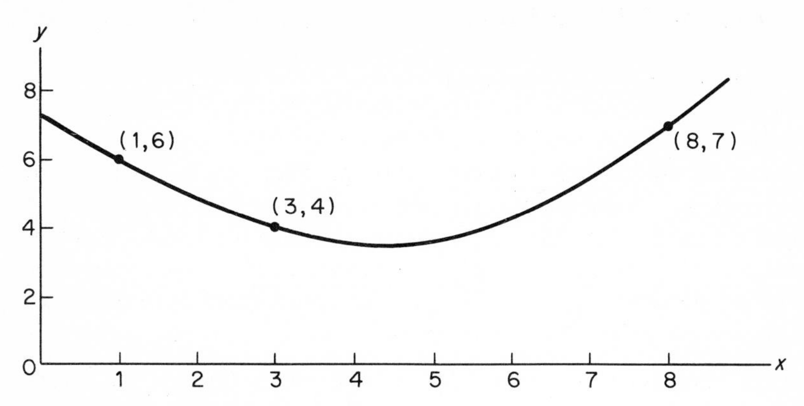

Figure A.2. Quadratic Function

or, alternatively,

$$6 = a + b + c$$
$$4 = a + 3b + 9c$$
$$7 = a + 8b + 64c.$$

These equations are solved simultaneously to give the coefficients a, b, and c; thus,

$$\begin{array}{rrrr} 6 = & a + & b + & c \\ -4 = & -a - & 3b - & 9c \\ \hline 2 = & & -2b - & 8c \end{array} \qquad \begin{array}{rrrr} 7 = & a + & 8b + & 64c \\ -4 = & -a - & 3b - & 9c \\ \hline 3 = & & 5b + & 55c \end{array}$$

By multiplying the equations by 5 and 2, we obtain

$$10 = -10b - 40c \quad \text{and} \quad 6 = 10b + 110c.$$

These two equations can be added to give

$$\begin{array}{rrr} 10 = & -10b - & 40c \\ 6 = & +10b + & 110c \\ \hline 16 = & + & 70c \end{array}$$

and

$$c = \frac{16}{70} = 0.228, \qquad b = \frac{-2 - 8c}{2} = \frac{-3.824}{2} = -1.912,$$
$$a = 6 - b - c = 6 + 1.912 - 0.228 = 7.684,$$

and the quadratic equation is

$$y = 7.684 - 1.912x + 0.228x^2.$$

This function is shown in Fig. A.2.

A.3 polynomial functions

The general form of the polynomial function is

$$y = a_0 + a_1x + a_2x^2 + \cdots + a_nx^n.$$

The method of establishing a function through data points, discussed in Secs. A.1 and A.2, also applies to the polynomial function of degree n. The procedure is to substitute the $(n + 1)$ data points into the general form of the equation and solve the resulting $(n + 1)$ equations simultaneously for the $(n + 1)$ parameters $a_0, a_1, a_2, \ldots, a_n$. The procedure for a cubic function is illustrated in the following example.

Example. Determine the cubic function that passes through the points (1, 2), (3, 6), (5, 7), and (7, 10). The four equations are

$$\begin{aligned} 2 &= a_0 + a_1(1) + a_2(1)^2 + a_3(1)^3 \\ 6 &= a_0 + a_1(3) + a_2(3)^2 + a_3(3)^3 \\ 7 &= a_0 + a_1(5) + a_2(5)^2 + a_3(5)^3 \\ 10 &= a_0 + a_1(7) + a_2(7)^2 + a_3(7)^3. \end{aligned}$$

The four equations are solved simultaneously to give $a_0 = -2.69, a_1 = 5.89$, $a_2 = -1.31$, and $a_3 = 0.10$. The resulting function is shown in Fig. A.3.

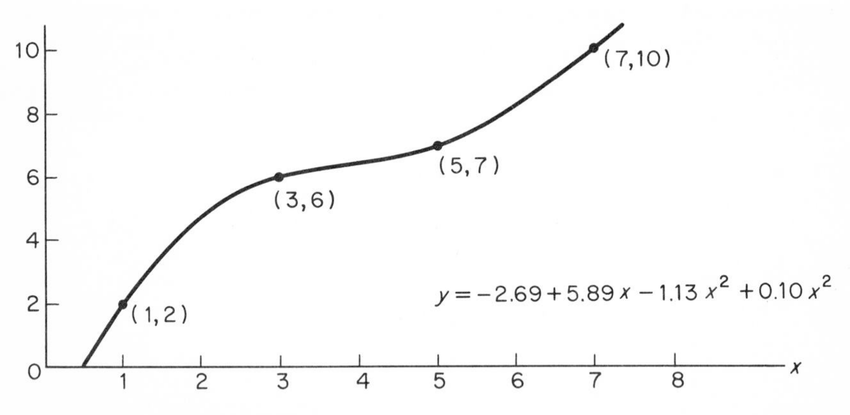

Figure A.3. Cubic Function

A.4 exponential functions

The exponential function has the general form $y = k(a)^{cx}$, where k and c are parameters and a is the base. An exponential function can be established through any two data points with the exception of $(x = 0, y = 0)$. The method of establishing the function is to write the exponential function as a logarithmic function, and solve the resulting two equations simultaneously for the parameters k and c. The procedure is illustrated by the following example.

Example. Determine the exponential function that passes through the data points (1, 2) and (4, 6). Use the base e. The general form of the function is

$$y = ke^{cx},$$

which when written in terms of logarithms becomes

$$\ln(y) = \ln(k) + cx.$$

Substituting the two data points gives the two equations

$$\ln(2) = \ln(k) + c(1)$$
$$\ln(6) = \ln(k) + c(4),$$

or, alternatively,

$$0.69315 = \ln(k) + c$$
$$1.79176 = \ln(k) + 4c.$$

Solving these two equations simultaneously gives $\ln(k) = 0.32695$ and $c = 0.36620$. The antilog of $\ln(k)$ is $k = 1.386$. The exponential function is thus

$$y = 1.386e^{0.36620x}.$$

This function is shown in Fig. A.4.

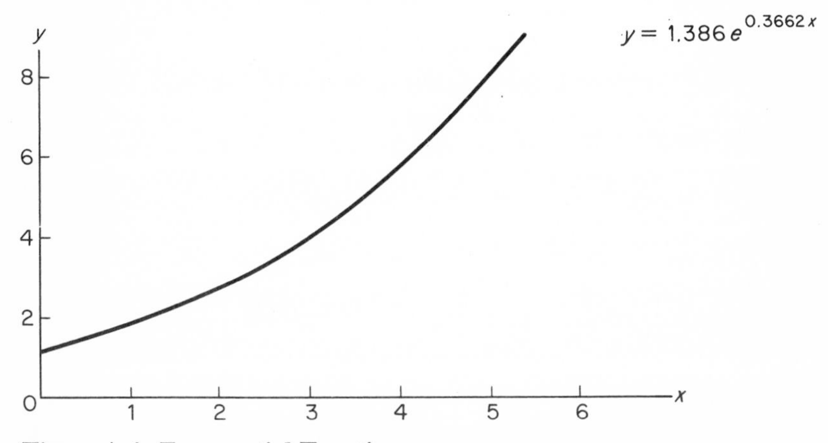

Figure A.4. Exponential Function

A.5 logarithmic functions

The general form of the logarithmic function is

$$y = a \ln (bx),$$

where x is the independent variable, y is the dependent variable, a and b are parameters, and the base of the logarithmic function is e. To determine the value of the parameters a and b, we make the following substitutions.

Let

$$y = \frac{1}{c} \ln \left(\frac{1}{k} x\right)$$

or

$$cy = \ln \left(\frac{1}{k} x\right).$$

The antilog of both sides (see Appendix B) gives

$$e^{cy} = \frac{1}{k} x$$

or

$$x = ke^{cy}.$$

We can now solve for the parameters k and c, using the method discussed in Sec. A.4. We must recognize that x is the dependent variable and y is the independent variable in the above transformation. The procedure is illustrated in the following example.

Example. Determine the logarithmic function that passes through the two data points (3, 1) and (6, 3).

The two equations are

$$\ln (3) = \ln (k) + c(1)$$
$$\ln (6) = \ln (k) + c(3).$$

Solving simultaneously gives

$$2c = \ln 6 - \ln(3), \quad \text{or} \quad c = 0.34658,$$

and

$$\ln (k) = 0.75203 \quad \text{and} \quad k = 2.121.$$

Since $a = 1/c$ and $b = 1/k$, the logarithmic function is

$$y = 2.885 \ln (0.4715x).$$

This function is shown in Fig. A.5.

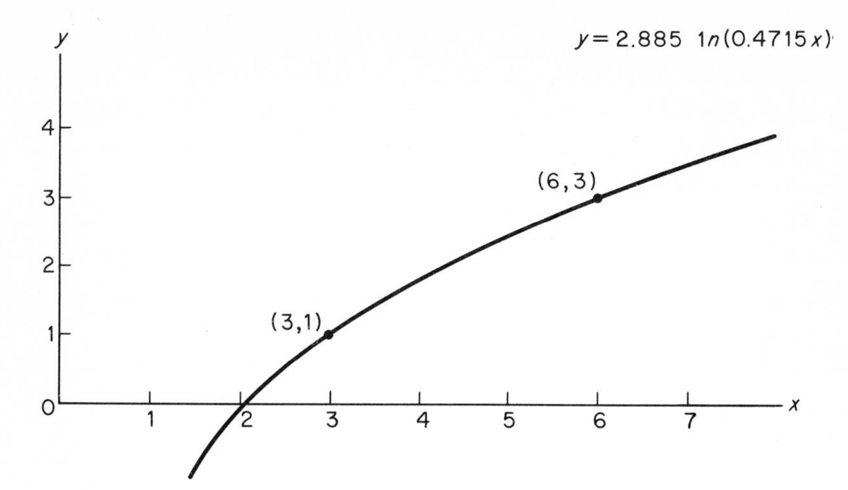

Figure A.5. Logarithmic Function

A.6 establishing functions through sets of data points

The method of least squares provides a technique for establishing a function through sets of data points. The technique was illustrated in Sec. 5.2 for linear, exponential, and power functions.

The method of least squares can also be used to establish polynomial functions through data points. We shall give the general formulation for establishing a pth-degree polynomial through a set of n data points. The procedure is illustrated for the 2nd-degree polynomial.

Consider the polynomial of the form

$$y = B_0 + B_1x + B_2x^2 + \cdots + B_px^p.$$

The method of least squares provides a technique for establishing a pth-degree polynomial through n data points ($n > p$). To determine the values of the parameters $B_0, B_1, B_2, \ldots, B_p$, we solve the following $p + 1$ equations simultaneously for the parameters.

$$\begin{aligned}
\sum y &= nB_0 + B_1 \sum x + \cdots + Bp \sum x^p \\
\sum xy &= B_0 \sum x + B_1 \sum x^2 + \cdots + Bp \sum x^{p+1} \\
\sum x^2y &= B_0 \sum x^2 + B_1 \sum x^3 + \cdots + Bp \sum x^{p+2} \\
&\vdots \\
\sum x^py &= B_0 \sum x^p + B_1 \sum x^{p+1} + \cdots + Bp \sum x^{2p}.
\end{aligned}$$

This procedure is illustrated by establishing a quadratic function through a set of data points in the following example.

Example. Establish a quadratic function through the data points: (1.0, 50), (1.5, 30), (2.0, 25), (2.5, 20), (3.0, 20), (4.0, 25), and (5.0, 50). These data points are plotted in Fig. A.6.

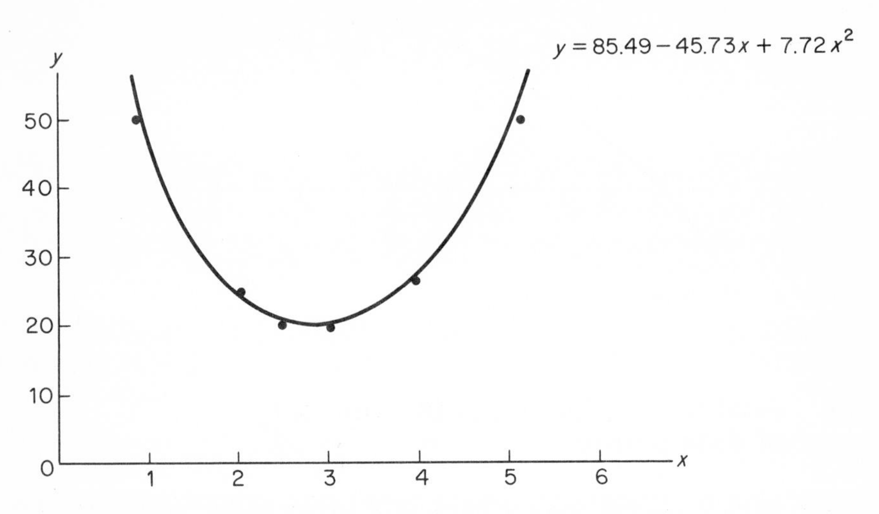

Figure A.6. Quadratic Function Established by Least Squares

A quadratic function is of degree 2. To determine the parameters B_0, B_1, and B_2 of the quadratic function

$$y = B_0 + B_1 x + B_2 x^2,$$

we solve the following three equations simultaneously:

$$\begin{aligned} \sum y &= nB_0 + B_1 \sum x + B_2 \sum x^2 \\ \sum xy &= B_0 \sum x + B_1 \sum x^2 + B_2 \sum x^3 \\ \sum x^2 y &= B_0 \sum x^2 + B_1 \sum x^3 + B_2 \sum x^4, \end{aligned}$$

where

	x	y	xy	x^2y	x^2	x^3	x^4
	1.0	50	50	50.0	1.0	1.00	1.00
	1.5	30	45	67.5	2.25	3.38	5.06
	2.0	25	50	100.0	4.00	8.00	16.00
	2.5	20	50	125.0	6.25	15.62	39.06
	3.0	20	60	180.0	9.00	27.00	81.00
	4.0	25	100	400.0	16.00	64.00	256.00
	5.0	50	250	1250.0	25.00	125.00	625.00
Σ	19.0	220	605	2172.5	63.50	244.00	1023.12

Inserting these summations into the three equations gives

$$220.0 = 7.0B_0 + 19.0B_1 + 63.5B_2$$
$$605.0 = 19.0B_0 + 63.5B_1 + 244.0B_2$$
$$2172.5 = 63.5B_0 + 244.0B_1 + 1023.1B_2.$$

Solving the equations simultaneously gives $B_0 = 85.49$, $B_1 = -45.73$, and $B_2 = 7.72$. The quadratic function is

$$y = 85.49 - 45.73x + 7.72x^2.$$

225

ndix B

ogarithms:

laws of exponents

B.1 logarithms as exponents

Consider the exponential function $y = a^x$, where y represents the value of the base number a raised to the exponent x. The definition of the logarithm of y is as follows: The logarithm of y to the base a is x. This is written as $\log_a (y) = x$. A *logarithm* is *an exponent*. To understand logarithms it is important to remember the simple fact that a logarithm is an exponent. In fact, it might be helpful for the student to consider the terms logarithm and exponent as synonymous. To illustrate the relationship between a number and its logarithm, consider the following examples.

Example. Determine $\log_{10} (100)$. The logarithm of 100 to the base 10 is the exponent to which 10 must be raised to give 100.
Since $100 = 10^2$, we know that $\log_{10} (100) = 2$.

Example. Determine $\log_{10} (10{,}000)$.
Since $10{,}000 = 10^4$, we conclude that $\log_{10} (10{,}000) = 4$.

Example. Determine $\log_2 (8)$.
Since $8 = 2^3$, we conclude that $\log_2 (8) = 3$.

Example. Determine $\log_6 (36)$.
Since $36 = 6^2$, we conclude that $\log_6 (36) = 2$.

Logarithms are used for two purposes in this text. These are (1) in the calculation of the product of two numbers, the quotient of two numbers, or a number raised to a power, and (2) the algebraic manipulation of exponential functions for the purpose of expressing the function in a linear form. To understand the use of logarithms, it will be useful to review the laws of exponents.

B.2 laws of exponents

Let a represent the base and m and n represent exponents. The three laws of exponents are

$$\text{(i)}\ a^m \cdot a^n = a^{m+n}$$

$$\text{(ii)}\ \frac{a^m}{a^n} = a^{m-n} = \begin{cases} a^{m-n} & \text{if } m > n \\ 1 & \text{if } m = n \\ \dfrac{1}{a^{n-m}} & \text{if } n > m \end{cases}$$

$$\text{(iii)}\ (a^m)^n = a^{mn}.$$

The usefulness of logarithms in calculations of products, quotients, and powers is based upon the laws of exponents. Numbers can be expressed with a common base. The product of two numbers is then equal to the sum of the exponents, the quotient of two numbers is equal to the difference between the exponents, and a number raised to a power is equal to the product of the exponents. These properties of exponents are illustrated by the following examples.

Example. Determine the product of 100 times 1000, using the laws of exponents.

Since $100 = 10^2$ and $1000 = 10^3$, we know that

$$(100)(1000) = (10^2)(10^3) = 10^{2+3} = 10^5 = 100{,}000.$$

Example. Determine the product of 16 times 64, using the laws of exponents.

Since $16 = 2^4$ and $64 = 2^6$, the product of 16 and 64 is

$$(2^4)(2^6) = 2^{4+6} = 2^{10} = 1024.$$

Example. Determine the quotient of 1,000,000 divided by 100.

Since $1{,}000{,}000 = 10^6$ and $100 = 10^2$, $1{,}000{,}000/100 = 10^6/10^2 = 10^{6-2} = 10^4 = 10{,}000$.

Example. Determine the value of 100 raised to the third power.

Since $100 = 10^2$, we use the third law of exponents to show that $(10^2)^3 = 10^6 = 1{,}000{,}000$.

B.3 laws of logarithms

The laws of exponents can be rewritten in terms of logarithms. If we represent any two nonzero, nonnegative numbers as x and y, then the laws of logarithms are

$$\text{(i)}\quad \log_a (xy) = \log_a (x) + \log_a (y)$$
$$\text{(ii)}\quad \log_a (x/y) = \log_a (x) - \log_a (y)$$
$$\text{(iii)}\quad \log_a (x^y) = y \log_a (x).$$

Logarithms are exponents of the base. To determine the value of the product, quotient, or power we must raise the base to the logarithm (or exponent). The process of raising the base to the logarithm is termed finding the antilog of the logarithm. In symbolic form we write

$$(x)(y) = \text{antilog}_a [\log_a (x) + \log_a (y)] = a^{\log_a(x)+\log_a(y)}.$$

The laws of logarithms are illustrated for the examples given in Sec. B.2.

Example. Determine the product of 100 times 1000, using logarithms.

$$\log_{10} (100)(1000) = \log_{10} (100) + \log_{10} (1000) = 2 + 3 = 5$$
$$\text{antilog}_{10} (5) = 10^5 = 100{,}000.$$

Example. Determine the product of 16 times 64, using logarithms.

$$\log_2 (16)(64) = \log_2 (16) + \log_2 (64) = 4 + 6 = 10$$
$$\text{antilog}_2 (10) = 2^{10} = 1024.$$

Example. Determine the quotient of 1,000,000 divided by 100, using logarithms.

$$\log_{10} (1{,}000{,}000/100) = \log_{10} (1{,}000{,}000) - \log_{10} (100) = 6 - 2 = 4$$
$$\text{antilog}_{10} (4) = 10^4 = 10{,}000.$$

Example. Determine the value of 100 raised to the third power, using logarithms.

$$\log_{10} (100)^3 = 3 \log_{10} (100) = 3(2) = 6$$
$$\text{antilog}_{10} (6) = 10^6 = 1{,}000{,}000.$$

B.4 determining logarithms

To determine the logarithm of a number, it is necessary to have a table that gives the logarithms of numbers using a specific base. Two bases are commonly used, the base 10 and the base e. Logarithms using the base 10 are termed common logarithms and are written as $\log (x)$. The absence of a subscript indicates that the base is 10. Logarithms using the base e are termed Naperian or natural logarithms and are written as ln (x). The symbol "ln" refers to the natural or Naperian logarithm with base e. Any base other than 10 or e is indicated by $\log_a(x)$. We shall first illustrate the common or base 10 logarithms.

A logarithm is composed of two parts, the characteristic and the mantissa. The method of determining the characteristic and the mantissa requires that the number be written in the form: $x.xxxx \cdot 10^n$. That is, the number for which the logarithm is to be obtained must be written in a form such that a simple integer is to the left of the decimal point with the remaining significant digits to the right of the decimal point. The number x.xxxx is termed the *argument*. The magnitude of the number is expressed as the argument multiplied by a power of 10. The characteristic of the logarithm is n, the power to which 10 is raised. We can illustrate the determination of the characteristic by the following examples.

Example. Determine the characteristic of 121.
Since $121 = 1.21 \times 10^2$, the characteristic is 2. Note also that the argument is 1.21.

Example. Determine the characteristic of 1640.
Since $1640 = 1.640 \times 10^3$, the characteristic is 3.

Example. Determine the characteristic of 0.0653.
Since $0.06535 = 6.53 \times 10^{-2}$, the characteristic is -2.

Example. Determine the characteristic of 7.54.
Since $7.54 = 7.54 \times 10^0$, the characteristic is 0.

The mantissa of the number is the exponent to which the base 10 must be raised to obtain the argument x.xxxx. Since the argument has a value between 1.0000 and 9.9999, the *mantissa must have a value between 0 and 1.* The relationship, again, between the base, the mantissa, and the argument is

$$(\text{base})^{\text{mantissa}} = \text{argument}.$$

If we include the characteristic, we see that

$$(\text{base})^{\text{mantissa}} \cdot (\text{base})^{\text{characteristic}} = \text{number}.$$

The logarithm of the number is thus the sum of the mantissa and the characteristic. Mantissas for common logarithms are given in Table III. As an example, we shall determine the logarithm of 297. We first note that $297 = 2.97 \times 10^2$. The characteristic of the logarithm is thus 2. The mantissa of the argument 2.97 is 0.4728. In terms of exponential notation, $297 = 10^{0.4728} \times 10^2$. The logarithm of the number is the exponent to which 10 must be raised to obtain 297. The logarithm of 297 is, therefore,

2.4728; i.e., log (297) = 2.4728. The method of determining logarithms is further illustrated by the following examples.

Example. log (121) = 2.0828, since $121 = 10^{0.0828} \times 10^{2}$.

Example. log (1640) = 3.2148, since $1640 = 10^{0.2148} \times 10^{3}$.

Example. log (0.0653) = −1.1851, since $0.0653 = 10^{0.8149} \times 10^{-2}$.

Example. log (7.54) = 0.8774, since $7.54 = 10^{0.8874} \times 10^{0}$.

B.5 using logarithms in calculations

The first law of logarithms specifies that the product of two numbers is given by the sum of their logarithms.

Example. Multiply 78 by 194.

$$\begin{aligned} \log 78 &= 1.8921 \\ \log 194 &= \underline{2.2878} \\ \text{Sum of logarithms} &= 4.1799 \\ \text{antilog}\,(4.1799) &= 15{,}130. \end{aligned}$$

The fourth digit from the left, 3, was obtained by linear interpolation.

The second law of logarithms states that the quotient of two numbers is given by the difference between two logarithms.

Example. Divide 657 by 132.

$$\begin{aligned} \log 657 &= 2.8041 \\ \log 132 &= \underline{2.1206} \\ \text{Difference of logarithms} &= 0.6835 \\ \text{antilog}\,(0.6835) &= 4.83. \end{aligned}$$

Example. Divide 456 by 3648.

$$\begin{aligned} \log 456 &= 2.6590 \\ \log 3648^{\dagger} &= \underline{3.5620} \\ \text{Difference of logarithms} &= -0.9030 \end{aligned}$$

This problem requires a slight modification. Negative exponents such as −0.9030 are not included in the table of mantissas. We can, however, rewrite the logarithm as

$$-0.9030 = +0.0970 - 1.0000.$$

†By interpolation.

Since logarithms are exponents, we actually have converted $10^{-0.9030}$ to $10^{0.0970} \cdot 10^{-1.0000}$. Thus, antilog (0.0970) = 1.25, and the quotient is $1.25 \cdot 10^{-1} = 0.125$.

The third law of logarithms states that to raise a number to a given power, the logarithm of the number is multiplied by the power.

Example. Raise 6.52 to the 3.25 power.

$$\begin{aligned} \log(6.52) &= 0.8142 \\ &\underline{\times 3.25} \\ \text{Product} &= 2.6462 \\ \text{antilog}(2.6462) &= 442.8. \end{aligned}$$

The third law of logarithms also applies in determining roots of a number.

Example. Evaluate $(174)^{0.21}$.

$$\begin{aligned} \log(174) &= 2.2405 \\ &\underline{\times 0.21} \\ \text{Product} &= 0.4705 \\ \text{antilog}(0.4705) &= 2.953. \end{aligned}$$

Table IV contains natural logarithms, i.e., logarithms with base e. The natural logarithms for numbers from 0 to 999 are given in this table. Both the characteristic and the mantissa are included in the table. Logarithms of numbers that exceed 999 can be determined by writing the number as a power of 10. Thus, the natural logarithm of 1230 is found by expressing 1230 as 123×10^1. The natural logarithm of 1230 is

$$\begin{aligned} \ln(1230) &= \ln(123) + \ln(10) \\ \ln(1230) &= 4.81218 + 2.30259 \\ \ln(1230) &= 7.11477. \end{aligned}$$

The laws of logarithms directly apply to natural logarithms.

B.6 manipulation of logarithmic and exponential functions

An exponential function may be expressed in terms of logarithms by expressing the terms of both sides of the equal sign in terms of logarithms. Thus,

$$y = kb^{f(x)}$$

can be expressed as

$$\log_a y = \log_a k + f(x) \log_a b.$$

Similarly, given the function

$$y = k \log_a f(x),$$

the function can be expressed in its exponential form as

$$a^y = f(x)^k.$$

A logarithm may be expressed with a different base by using the following relationship.

$$\log_a x = (\log_b x)(\log_a b).$$

Example. Convert $\log_{10}(25)$ to $\log_e(25)$.

$$\log_e(25) = \log_{10}(25)\log_e(10)$$
$$\log_e(25) = (1.3979)(2.30259) = 3.21888.$$

It is also helpful to recognize that

$$\log_a(a) = 1$$

and

$$\log_a(1) = 0.$$

tables

TABLE I. VALUES OF $(1 + i)^n$

n	$i = 0.005$	$i = 0.0075$	$i = 0.01$	$i = 0.0125$	$i = 0.015$	$i = 0.02$	$i = 0.03$	$i = 0.04$	$i = 0.05$	$i = 0.06$	$i = 0.07$	$i = 0.08$
1	1.005000	1.007500	1.010000	1.012500	1.015000	1.020000	1.030000	1.040000	1.050000	1.060000	1.070000	1.080000
2	1.010025	1.015056	1.020100	1.025156	1.030225	1.040400	1.060900	1.081600	1.102500	1.123600	1.144900	1.166400
3	1.015075	1.022669	1.030301	1.037971	1.045678	1.061208	1.092727	1.124864	1.157625	1.191016	1.225043	1.259712
4	1.020150	1.030339	1.040604	1.050945	1.061364	1.082432	1.125509	1.169859	1.215506	1.262477	1.310796	1.360489
5	1.025251	1.038067	1.050110	1.064082	1.077284	1.104081	1.159274	1.216653	1.276282	1.338226	1.402552	1.469328
6	1.030378	1.045852	1.061520	1.077383	1.093443	1.126162	1.194052	1.265319	1.340096	1.418519	1.500730	1.586874
7	1.035529	1.053696	1.072135	1.090850	1.109845	1.148686	1.229874	1.315932	1.407100	1.503630	1.605781	1.713824
8	1.040707	1.061599	1.082857	1.104486	1.126493	1.171659	1.266770	1.368569	1.477455	1.593848	1.718186	1.850930
9	1.045911	1.069561	1.093685	1.118292	1.143390	1.195093	1.304773	1.423312	1.551328	1.689479	1.838459	1.999004
10	1.051140	1.077583	1.104622	1.132271	1.160541	1.218994	1.343916	1.480244	1.628895	1.790848	1.967151	2.158925
11	1.056396	1.085664	1.115668	1.146424	1.177949	1.243374	1.384234	1.539454	1.710339	1.898299	2.104852	2.331629
12	1.061678	1.093807	1.126825	1.160755	1.195618	1.268242	1.425761	1.601032	1.795856	2.012196	2.252192	2.518170
13	1.066986	1.102010	1.138093	1.175264	1.213552	1.293607	1.468534	1.665074	1.885649	2.132928	2.404845	2.719624
14	1.072321	1.110276	1.149474	1.189955	1.231756	1.319479	1.512590	1.731676	1.979932	2.260904	2.578534	2.937194
15	1.077683	1.118603	1.160969	1.204829	1.250232	1.345868	1.557967	1.800944	2.078928	2.396558	2.759032	3.172169
16	1.083071	1.126992	1.172579	1.219890	1.268986	1.372786	1.604706	1.872981	2.182875	2.540352	2.952164	3.425943
17	1.088487	1.135445	1.184304	1.235138	1.288020	1.400241	1.652848	1.947900	2.292018	2.692773	3.158815	3.700018
18	1.093929	1.143960	1.196147	1.250577	1.307341	1.428246	1.702433	2.025817	2.406619	2.854339	3.379932	3.996019
19	1.099399	1.152540	1.208109	1.266210	1.326951	1.456811	1.753506	2.106849	2.526950	3.025600	3.616527	4.315701
20	1.104896	1.161184	1.220190	1.282037	1.346855	1.485947	1.806111	2.191123	2.653298	3.207135	3.869684	4.660957
21	1.110420	1.169893	1.232392	1.298063	1.367058	1.515666	1.860295	2.278768	2.785963	3.399564	4.140562	5.033834
22	1.115972	1.178667	1.244716	1.314288	1.387564	1.545980	1.916103	2.369919	2.925261	3.603537	4.430402	5.436540
23	1.121552	1.187507	1.257163	1.330717	1.408377	1.576899	1.973587	2.464716	3.071524	3.819750	4.740530	5.871464
24	1.127160	1.196414	1.269735	1.347351	1.429503	1.608437	2.032794	2.563304	3.225100	4.048935	5.072367	6.341181
25	1.132796	1.205387	1.282432	1.364193	1.450945	1.640606	2.093778	2.665836	3.386355	4.291871	5.427433	6.848475

26	1.138460	1.214427	1.295256	1.381245	1.472710	1.673418	2.156591	2.772470	3.555673	4.549383	5.807353	7.396353
27	1.144152	1.223535	1.308209	1.398511	1.494800	1.706886	2.221289	2.883369	3.733456	4.822346	6.213868	7.988061
28	1.149873	1.232712	1.321291	1.415992	1.517222	1.741024	2.287928	2.998703	3.920129	5.111687	6.648838	8.627106
29	1.155622	1.241957	1.334504	1.433692	1.539981	1.775845	2.356566	3.118651	4.116136	5.418388	7.114257	9.317275
30	1.161400	1.251272	1.347849	1.451613	1.563080	1.811362	2.427263	3.243398	4.321942	5.743491	7.612255	10.062657
31	1.167207	1.260656	1.361327	1.469759	1.586526	1.847589	2.500080	3.373133	4.538039	6.088101	8.145113	10.867669
32	1.173043	1.270111	1.374941	1.488131	1.610324	1.884541	2.575083	3.508059	4.764941	6.453387	8.715271	11.737083
33	1.178908	1.279637	1.388690	1.506732	1.634479	1.922231	2.652335	3.648381	5.003186	6.840590	9.325340	12.676050
34	1.184803	1.289234	1.402577	1.525566	1.658996	1.960676	2.731905	3.794316	5.253348	7.251025	9.978114	13.690134
35	1.190727	1.298904	1.416603	1.544636	1.683881	1.999890	2.813862	3.946089	5.516015	7.686087	10.676581	14.785344
36	1.196681	1.308645	1.430769	1.563944	1.709140	2.039887	2.898278	4.103933	5.791816	8.147252	11.423942	15.968172
37	1.202664	1.318460	1.445076	1.583493	1.734777	2.080685	2.985227	4.268090	6.081407	8.636087	12.223618	17.245626
38	1.208677	1.328349	1.459527	1.603287	1.760798	2.122299	3.074783	4.438813	6.385477	9.154252	13.079271	18.625276
39	1.214721	1.338311	1.474123	1.623328	1.787210	2.164745	3.167027	4.616366	6.704751	9.703507	13.994820	20.115298
40	1.220794	1.348349	1.488864	1.643619	1.814018	2.208040	3.262038	4.801021	7.039989	10.285718	14.974458	21.724522
41	1.226898	1.358461	1.503752	1.664165	1.841229	2.252200	3.359899	4.993061	7.391988	10.902861	16.022670	23.462483
42	1.233033	1.368650	1.518790	1.684967	1.868847	2.297244	3.460696	5.192784	7.761588	11.557033	17.144257	25.339482
43	1.239198	1.378915	1.533978	1.706029	1.896880	2.343189	3.564517	5.400495	8.149667	12.250455	18.344355	27.366640
44	1.245394	1.389256	1.549318	1.727354	1.925333	2.390053	3.671452	5.616515	8.557150	12.985482	19.628460	29.555972
45	1.251621	1.399676	1.564811	1.748946	1.954213	2.437854	3.781596	5.841176	8.985008	13.764611	21.002452	31.920449
46	1.257879	1.410173	1.580459	1.770808	1.983526	2.486611	3.895044	6.074823	9.434258	14.590487	22.472623	34.474085
47	1.264168	1.420750	1.596263	1.792943	2.013279	2.536344	4.011895	6.317816	9.905971	15.465917	24.045707	37.232012
48	1.270489	1.431405	1.612226	1.815355	2.043478	2.587070	4.132252	6.570528	10.401270	16.393872	25.728907	40.210573
49	1.276842	1.442141	1.628348	1.838047	2.074130	2.638812	4.256219	6.833349	10.921333	17.377504	27.529930	43.427420
50	1.283226	1.452957	1.644632	1.861022	2.105242	2.691588	4.383906	7.106683	11.467400	18.420154	29.457025	46.901613
51	1.289642	1.463854	1.661078	1.884285	2.136821	2.745420	4.515423	7.390951	12.040770	19.525364	31.519017	50.653742
52	1.296090	1.474833	1.677689	1.907839	2.168873	2.800328	4.650886	7.686589	12.642808	20.696885	33.725348	54.706041
53	1.302571	1.485894	1.694466	1.931687	2.201406	2.856335	4.790412	7.994052	13.274949	21.938699	36.086122	59.082524
54	1.309083	1.497038	1.711410	1.955833	2.234428	2.913461	4.934125	8.313814	13.938696	23.255020	38.612151	63.809126
55	1.315629	1.508266	1.728525	1.980281	2.267944	2.971731	5.082149	8.646367	14.635631	24.650322	41.315002	68.913856
56	1.322207	1.519578	1.745810	2.005034	2.301963	3.031165	5.234613	8.992222	15.367413	26.129341	44.207052	74.426965
57	1.328818	1.530975	1.763268	2.030097	2.336493	3.091789	5.391651	9.351911	16.133783	27.697101	47.301545	80.381122
58	1.335462	1.542457	1.780901	2.055473	2.371540	3.153624	5.553401	9.725987	16.942572	29.358927	50.612653	86.811612
59	1.342139	1.554026	1.798710	2.081167	2.407113	3.216697	5.720003	10.115026	17.789701	31.120463	54.155539	93.756540
60	1.348850	1.565681	1.816697	2.107181	2.443220	3.281031	5.891603	10.519627	18.679186	32.987691	57.946427	101.257064

TABLE II. VALUES OF $(1 + i)^{-n}$

n	$i = 0.005$	$i = 0.0075$	$i = 0.01$	$i = 0.0125$	$i = 0.015$	$i = 0.02$	$i = 0.03$	$i = 0.04$	$i = 0.05$	$i = 0.06$	$i = 0.07$	$i = 0.08$
1	0.995025	0.992556	0.990099	0.987654	0.985222	0.980392	0.970874	0.961538	0.952381	0.943396	0.934779	0.925926
2	0.990074	0.985167	0.980296	0.975461	0.970662	0.961169	0.942596	0.924556	0.907029	0.889996	0.873439	0.857339
3	0.985149	0.977833	0.970590	0.963418	0.956317	0.942322	0.915142	0.888996	0.863838	0.839619	0.816298	0.793832
4	0.980248	0.970554	0.960980	0.951524	0.942184	0.923845	0.888487	0.854804	0.822702	0.792094	0.762895	0.735030
5	0.975371	0.963329	0.951466	0.939777	0.928260	0.905731	0.862609	0.821927	0.783526	0.747258	0.712986	0.680583
6	0.970518	0.956158	0.942045	0.928175	0.914542	0.887971	0.837484	0.790315	0.746215	0.704961	0.666342	0.630170
7	0.965690	0.949040	0.932718	0.916716	0.901027	0.870560	0.813092	0.759918	0.710681	0.665057	0.622750	0.583490
8	0.960885	0.941975	0.923483	0.905398	0.887711	0.853490	0.789409	0.730690	0.676839	0.627412	0.582009	0.540269
9	0.956105	0.934963	0.914340	0.894221	0.874592	0.836755	0.766417	0.702587	0.644609	0.591898	0.543934	0.500249
10	0.951348	0.928003	0.905287	0.883181	0.861667	0.820348	0.744094	0.675564	0.613913	0.558395	0.508349	0.463193
11	0.946615	0.921095	0.896324	0.872277	0.848933	0.804263	0.722421	0.649581	0.584679	0.526788	0.475093	0.428883
12	0.941905	0.914238	0.887449	0.861509	0.836387	0.778493	0.701380	0.624597	0.556837	0.496969	0.444012	0.397114
13	0.937219	0.907432	0.878663	0.850873	0.824027	0.773033	0.680951	0.600574	0.530321	0.468839	0.414964	0.367698
14	0.932556	0.900677	0.869963	0.840368	0.811849	0.757875	0.661118	0.577475	0.505068	0.442301	0.387817	0.340461
15	0.927917	0.893973	0.861349	0.829993	0.799852	0.743015	0.641862	0.555264	0.481017	0.417265	0.362446	0.315243
16	0.923300	0.887318	0.852821	0.819746	0.788031	0.728446	0.623167	0.533908	0.458112	0.393646	0.338735	0.291890
17	0.918707	0.880712	0.844377	0.809626	0.776385	0.714163	0.605016	0.513373	0.436297	0.371364	0.316574	0.270269
18	0.914136	0.874156	0.836017	0.799631	0.764912	0.700159	0.587395	0.493628	0.415521	0.350344	0.295864	0.250249
19	0.909588	0.867649	0.827740	0.789759	0.753607	0.686431	0.570286	0.474642	0.395734	0.330513	0.276508	0.231712
20	0.905063	0.861190	0.819544	0.780009	0.742470	0.672971	0.553676	0.456387	0.376889	0.311805	0.258419	0.214548
21	0.900560	0.854779	0.811430	0.770379	0.731498	0.659776	0.537549	0.438834	0.358942	0.294155	0.241513	0.198656
22	0.896080	0.848416	0.803396	0.760868	0.720688	0.646839	0.521892	0.421955	0.341850	0.277505	0.225713	0.183941
23	0.891622	0.842100	0.795442	0.751475	0.710037	0.634156	0.506692	0.405726	0.325571	0.261797	0.210947	0.170315
24	0.887186	0.835831	0.787566	0.742197	0.699544	0.621721	0.491934	0.390121	0.310068	0.246979	0.197147	0.157699
25	0.882772	0.829609	0.779768	0.733034	0.689206	0.609531	0.477606	0.375117	0.295303	0.232999	0.184249	0.146018

26	0.878380	0.823434	0.772048	0.723984	0.679021	0.597579	0.463695	0.360689	0.281241	0.219810	0.172195	0.135202
27	0.874010	0.817304	0.764404	0.715046	0.668986	0.585862	0.450189	0.346817	0.267848	0.207368	0.160930	0.125187
28	0.869662	0.811220	0.756836	0.706219	0.659099	0.574375	0.437077	0.333477	0.255094	0.195630	0.150402	0.115914
29	0.865335	0.805181	0.749342	0.697500	0.649359	0.563112	0.424346	0.320651	0.242946	0.184557	0.140563	0.107328
30	0.861030	0.799187	0.741923	0.688889	0.639762	0.552071	0.411987	0.308319	0.231377	0.174110	0.131367	0.099377
31	0.856746	0.793238	0.734577	0.680384	0.630308	0.541246	0.399987	0.296460	0.220359	0.164255	0.122773	0.092016
32	0.852484	0.787333	0.727304	0.671984	0.620993	0.530633	0.388337	0.285058	0.209866	0.154957	0.114741	0.085200
33	0.848242	0.781472	0.720103	0.663688	0.611816	0.520229	0.377026	0.274094	0.199873	0.146186	0.107235	0.078889
34	0.844022	0.775654	0.712973	0.655494	0.602774	0.510028	0.366045	0.263552	0.190355	0.137912	0.100219	0.073045
35	0.839823	0.769880	0.705914	0.647402	0.593866	0.500028	0.355383	0.253415	0.181290	0.130105	0.093663	0.067635
36	0.835645	0.764149	0.698925	0.639409	0.585090	0.490223	0.345032	0.243669	0.172657	0.122741	0.087535	0.062635
37	0.831487	0.758461	0.692005	0.631515	0.576443	0.480611	0.334983	0.234297	0.164436	0.115793	0.081809	0.057986
38	0.827351	0.752814	0.685153	0.623719	0.567924	0.471187	0.325226	0.225285	0.156605	0.109239	0.076457	0.053690
39	0.823235	0.747210	0.678370	0.616018	0.559531	0.461948	0.315754	0.216621	0.149148	0.103056	0.071455	0.049713
40	0.819139	0.741648	0.671653	0.608413	0.551262	0.452890	0.306557	0.208289	0.142046	0.097222	0.066780	0.046031
41	0.815064	0.736127	0.665003	0.600902	0.543116	0.444010	0.297628	0.200278	0.135282	0.091719	0.062412	0.042621
42	0.811008	0.730647	0.658419	0.593484	0.535089	0.435304	0.288959	0.192575	0.128840	0.086527	0.058329	0.039464
43	0.806974	0.725208	0.651900	0.586157	0.527182	0.426769	0.280543	0.185168	0.122704	0.081630	0.054513	0.036541
44	0.802959	0.719810	0.645445	0.578920	0.519391	0.418401	0.272372	0.178046	0.116861	0.077009	0.050946	0.033834
45	0.798964	0.714451	0.639055	0.571773	0.511715	0.410197	0.264439	0.171198	0.111297	0.072650	0.047613	0.031329
46	0.794989	0.709133	0.632728	0.564714	0.504153	0.402154	0.256737	0.164614	0.105997	0.068538	0.044499	0.029007
47	0.791034	0.703854	0.626463	0.557742	0.496702	0.394268	0.249259	0.158283	0.100949	0.064658	0.041587	0.026859
48	0.787098	0.698614	0.620260	0.550856	0.489362	0.386538	0.241999	0.152195	0.096142	0.060998	0.038867	0.024869
49	0.783182	0.693414	0.614119	0.544056	0.482130	0.378958	0.234950	0.146341	0.091564	0.057546	0.036324	0.023027
50	0.779286	0.688252	0.608039	0.537339	0.475005	0.371528	0.228107	0.140713	0.087204	0.054288	0.033948	0.021321
51	0.775409	0.683128	0.602019	0.530705	0.467985	0.364243	0.221463	0.135301	0.083051	0.051215	0.031727	0.019742
52	0.771551	0.678043	0.596058	0.524153	0.461069	0.357101	0.215013	0.130097	0.079096	0.048316	0.029651	0.018280
53	0.767713	0.672995	0.590156	0.517682	0.454255	0.350100	0.208750	0.125093	0.075330	0.045582	0.027712	0.016926
54	0.763893	0.667986	0.584313	0.511291	0.447542	0.343234	0.202670	0.120282	0.071743	0.043002	0.025899	0.015672
55	0.760093	0.663013	0.578528	0.504979	0.440928	0.336504	0.196767	0.115656	0.068326	0.040567	0.024204	0.014511
56	0.756311	0.658077	0.572800	0.498745	0.434412	0.329906	0.191036	0.111207	0.065073	0.038271	0.022621	0.013436
57	0.752548	0.653178	0.567129	0.492587	0.427992	0.323437	0.185472	0.106930	0.061974	0.036105	0.021141	0.012441
58	0.748804	0.648316	0.561514	0.486506	0.421667	0.317095	0.180070	0.102817	0.059023	0.034061	0.019758	0.011519
59	0.745079	0.643490	0.555954	0.480500	0.415435	0.310878	0.174826	0.098863	0.056213	0.032133	0.018465	0.010666
60	0.741372	0.638700	0.550450	0.474568	0.409296	0.304782	0.169733	0.095060	0.053536	0.030314	0.017257	0.009876

TABLE III. COMMON LOGARITHMS

N	0	1	2	3	4	5	6	7	8	9
10	0000	0043	0086	0128	0170	0212	0253	0294	0334	0374
11	0414	0453	0492	0531	0569	0607	0645	0682	0719	0755
12	0792	0828	0864	0899	0934	0969	1004	1038	1072	1106
13	1139	1173	1206	1239	1271	1303	1335	1367	1399	1430
14	1461	1492	1523	1553	1584	1614	1644	1673	1703	1732
15	1761	1790	1818	1847	1875	1903	1931	1959	1987	2014
16	2041	2068	2095	2122	2148	2175	2201	2227	2253	2279
17	2304	2330	2355	2380	2405	2430	2455	2480	2504	2529
18	2553	2577	2601	2625	2648	2672	2695	2718	2742	2765
19	2788	2810	2833	2856	2878	2900	2923	2945	2967	2989
20	3010	3032	3054	3075	3096	3118	3139	3160	3181	3201
21	3222	3243	3263	3284	3304	3324	3345	3365	3385	3404
22	3424	3444	3464	3483	3502	3522	3541	3560	3579	3598
23	3617	3636	3655	3674	3692	3711	3729	3747	3766	3784
24	3802	3820	3838	3856	3874	3892	3909	3927	3945	3962
25	3979	3997	4014	4031	4048	4065	4082	4099	4116	4133
26	4150	4166	4183	4200	4216	4232	4249	4265	4281	4298
27	4314	4330	4346	4362	4378	4393	4409	4425	4440	4456
28	4472	4487	4502	4518	4533	4548	4564	4579	4594	4609
29	4624	4639	4654	4669	4683	4698	4713	4728	4742	4757
30	4771	4786	4800	4814	4829	4843	4857	4871	4886	4900
31	4914	4928	4942	4955	4969	4983	4997	5011	5024	5038
32	5051	5065	5079	5092	5105	5119	5132	5145	5159	5172
33	5185	5198	5211	5224	5237	5250	5263	5276	5289	5302
34	5315	5328	5340	5353	5366	5378	5391	5403	5416	5428
35	5441	5453	5465	5478	5490	5502	5514	5527	5539	5551
36	5563	5575	5587	5599	5611	5623	5635	5647	5658	5670
37	5682	5694	5705	5717	5729	5740	5752	5763	5775	5786
38	5798	5809	5821	5832	5843	5855	5866	5877	5888	5899
39	5911	5922	5933	5944	5955	5966	5977	5988	5999	6010
40	6021	6031	6042	6053	6064	6075	6085	6096	6107	6117
41	6128	6138	6149	6160	6170	6180	6191	6201	6212	6222
42	6232	6243	6253	6263	6274	6284	6294	6304	6314	6325
43	6335	6345	6355	6365	6375	6385	6395	6405	6415	6425
44	6435	6444	6454	6464	6474	6484	6493	6503	6513	6522
45	6532	6542	6551	6561	6571	6580	6590	6599	6609	6618
46	6628	6637	6646	6656	6665	6675	6684	6693	6702	6712
47	6721	6730	6739	6749	6758	6767	6776	6785	6794	6803
48	6812	6821	6830	6839	6848	6857	6866	6875	6884	6893
49	6902	6911	6920	6928	6937	6946	6955	6964	6972	6981
50	6990	6998	7007	7016	7024	7033	7042	7050	7059	7067
51	7076	7084	7093	7101	7110	7118	7126	7135	7143	7152
52	7160	7168	7177	7185	7193	7202	7210	7218	7226	7235
53	7243	7251	7259	7267	7275	7284	7292	7300	7308	7316
54	7324	7332	7340	7348	7356	7364	7372	7380	7388	7396

TABLE III. COMMON LOGARITHMS (Continued)

N	0	1	2	3	4	5	6	7	8	9
55	7404	7412	7419	7427	7435	7443	7451	7459	7466	7474
56	7482	7490	7497	7505	7513	7520	7528	7536	7543	7551
57	7559	7566	7574	7582	7589	7597	7604	7612	7619	7627
58	7634	7642	7649	7657	7664	7672	7679	7686	7694	7701
59	7709	7716	7723	7731	7738	7745	7752	7760	7767	7774
60	7782	7789	7796	7803	7810	7818	7825	7832	7839	7846
61	7853	7860	7868	7875	7882	7889	7896	7903	7910	7917
62	7924	7931	7938	7945	7952	7959	7966	7973	7980	7987
63	7993	8000	8007	8014	8021	8028	8035	8041	8048	8055
64	8062	8069	8075	8082	8089	8096	8102	8109	8116	8122
65	8129	8136	8142	8149	8156	8162	8169	8176	8182	8189
66	8195	8202	8209	8215	8222	8228	8235	8241	8248	8254
67	8261	8267	8274	8280	8287	8293	8299	8306	8312	8319
68	8325	8331	8338	8344	8351	8357	8363	8370	8376	8382
69	8388	8395	8401	8407	8414	8420	8426	8432	8439	8445
70	8451	8457	8463	8470	8476	8482	8488	8494	8500	8506
71	8513	8519	8525	8531	8537	8543	8549	8555	8561	8567
72	8573	8579	8585	8591	8597	8603	8609	8615	8621	8627
73	8633	8639	8645	8651	8657	8663	8669	8675	8681	8686
74	8692	8698	8704	8710	8716	8722	8727	8733	8739	8745
75	8751	8756	8762	8768	8774	8779	8785	8791	8797	8802
76	8808	8814	8820	8825	8831	8837	8842	8848	8854	8859
77	8865	8871	8876	8882	8887	8893	8899	8904	8910	8915
78	8921	8927	8932	8938	8943	8949	8954	8960	8965	8971
79	8976	8982	8987	8993	8998	9004	9009	9015	9020	9025
80	9031	9036	9042	9047	9053	9058	9063	9069	9074	9079
81	9085	9090	9096	9101	9106	9112	9117	9122	9128	9133
82	9138	9143	9149	9154	9159	9165	9170	9175	9180	9186
83	9191	9196	9201	9206	9212	9217	9222	9227	9232	9238
84	9243	9248	9253	9258	9263	9269	9274	9279	9284	9289
85	9294	9299	9304	9309	9315	9320	9325	9330	9335	9340
86	9345	9350	9355	9360	9365	9370	9375	9380	9385	9390
87	9395	9400	9405	9410	9415	9420	9425	9430	9435	9440
88	9445	9450	9455	9460	9465	9469	9474	9479	9484	9489
89	9494	9499	9504	9509	9513	9518	9523	9528	9533	9538
90	9542	9547	9552	9557	9562	9566	9571	9576	9581	9586
91	9590	9595	9600	9605	9609	9614	9619	9624	9628	9633
92	9638	9643	9647	9652	9657	9661	9666	9671	9675	9680
93	9685	9689	9694	9699	9703	9708	9713	9717	9722	9727
94	9731	9736	9741	9745	9750	9754	9759	9763	9768	9773
95	9777	9782	9786	9791	9795	9800	9805	9809	9814	9818
96	9823	9827	9832	9836	9841	9845	9850	9854	9859	9863
97	9868	9872	9877	9881	9886	9890	9894	9899	9903	9908
98	9912	9917	9921	9926	9930	9934	9939	9943	9948	9952
99	9956	9961	9965	9969	9974	9978	9983	9987	9991	9996

This table reprinted with permission from *Standard Mathematical Tables*, 18th edition, The Chemical Rubber Co.

TABLE IV. NATURAL OR NAPERIAN LOGARITHMS

0.000–0.499

N	0	1	2	3	4	5	6	7	8	9
0.00	−∞	−6‡ .90776	−6 .21461	−5 .80914	−5 .52146	−5 .29832	−5 .11000	−4 .96185	−4 .82831	−4 .71053
.01	−4.60517	.50986	.42285	.34281	.26870	.19971	.13517	.07454	.01738	*.96332
.02	−3.91202	.86323	.81671	.77226	.72970	.68888	.64966	.61192	.57555	.54046
.03	.50656	.47377	.44202	.41125	.38139	.35241	.32424	.29684	.27017	.24419
.04	.21888	.19418	.17009	.14656	.12357	.10109	.07911	.05761	.03655	.01593
.05	−2.99573	.97593	.95651	.93746	.91877	.90042	.88240	.86470	.84731	.83022
.06	.81341	.79688	.78062	.76462	.74887	.73337	.71810	.70306	.68825	.67365
.07	.65926	.64508	.63109	.61730	.60369	.59027	.57702	.56395	.55105	.53831
.08	.52573	.51331	.50104	.48891	.47694	.46510	.45341	.44185	.43042	.41912
.09	.40795	.39690	.38597	.37516	.36446	.35388	.34341	.33304	.32279	.31264
0.10	−2.30259	.29263	.28278	.27303	.26336	.25379	.24432	.23493	.22562	.21641
.11	.20727	.19823	.18926	.18037	.17156	.16282	.15417	.14558	.13707	.12863
.12	.12026	.11196	.10373	.09557	.08747	.07944	.07147	.06357	.05573	.04794
.13	.04022	.03256	.02495	.01741	.00992	.00248	*.99510	*.98777	*.98050	*.97328
.14	−1.96611	.95900	.95193	.94491	.93794	.93102	.92415	.91732	.91054	.90381
.15	.89712	.89048	.88387	.87732	.87080	.86433	.85790	.85151	.84516	.83885
.16	.83258	.82635	.82016	.81401	.80789	.80181	.79577	.78976	.78379	.77786
.17	.77196	.76609	.76026	.75446	.74870	.74297	.73727	.73161	.72597	.72037
.18	.71480	.70926	.70375	.69827	.69282	.68740	.68201	.67665	.67131	.66601
.19	.66073	.65548	.65026	.64507	.63990	.63476	.62964	.62455	.61949	.61445
0.20	−1.60944	.60445	.59949	.59455	.58964	.58475	.57988	.57504	.57022	.56542
.21	.56065	.55590	.55117	.54646	.54178	.53712	.53248	.52786	.52326	.51868
.22	.51413	.50959	.50508	.50058	.49611	.49165	.48722	.48281	.47841	.47403
.23	.46968	.46534	.46102	.45672	.45243	.44817	.44392	.43970	.43548	.43129
.24	.42712	.42296	.41882	.41469	.41059	.40650	.40242	.39837	.39433	.39030
.25	.38629	.38230	.37833	.37437	.37042	.36649	.36258	.35868	.35480	.35093
.26	.34707	.34323	.33941	.33560	.33181	.32803	.32426	.32051	.31677	.31304
.27	.30933	.30564	.30195	.29828	.29463	.29098	.28735	.28374	.28013	.27654
.28	.27297	.26940	.26585	.26231	.25878	.25527	.25176	.24827	.24479	.24133
.29	.23787	.23443	.23100	.22758	.22418	.22078	.21740	.21402	.21066	.20731
0.30	−1.20397	.20065	.19733	.19402	.19073	.18744	.18417	.18091	.17766	.17441
.31	.17118	.16796	.16475	.16155	.15836	.15518	.15201	.14885	.14570	.14256
.32	.13943	.13631	.13320	.13010	.12701	.12393	.12086	.11780	.11474	.11170
.33	.10866	.10564	.10262	.09961	.09661	.09362	.09064	.08767	.08471	.08176
.34	.07881	.07587	.07294	.07002	.06711	.06421	.06132	.05843	.05555	.05268
.35	−1.04982	.04697	.04412	.04129	.03846	.03564	.03282	.03002	.02722	.02443
.36	.02165	.01888	.01611	.01335	.01060	.00786	.00512	.00239	*.99967	*.99696
.37	−0.99425	.99155	.98886	.98618	.98350	.98083	.97817	.97551	.97286	.97022
.38	.96758	.96496	.96233	.95972	.95711	.95451	.95192	.94933	.94675	.94418
.39	.94161	.93905	.93649	.93395	.93140	.92887	.92634	.92382	.92130	.91879
0.40	−0.91629	.91379	.91130	.90882	.90634	.90387	.90140	.89894	.89649	.89404
.41	.89160	.88916	.88673	.88431	.88189	.87948	.87707	.87467	.87227	.86988
.42	.86750	.86512	.86275	.86038	.85802	.85567	.85332	.85097	.84863	.84630
.43	.84397	.84165	.83933	.83702	.83471	.83241	.83011	.82782	.82554	.82326
.44	.82098	.81871	.81645	.81419	.81193	.80968	.80744	.80520	.80296	.80073
.45	.79851	.79629	.79407	.79186	.78966	.78746	.78526	.78307	.78089	.77871
.46	.77653	.77436	.77219	.77003	.76787	.76572	.76357	.76143	.75929	.75715
.47	.75502	.75290	.75078	.74866	.74655	.74444	.74234	.74024	.73814	.73605
.48	.73397	.73189	.72981	.72774	.72567	.72361	.72155	.71949	.71744	.71539
.49	.71335	.71131	.70928	.70725	.70522	.70320	.70118	.69917	.69716	.69515

TABLE IV. NATURAL OR NAPERIAN LOGARITHMS (Continued)
0.500–0.999

N	0	1	2	3	4	5	6	7	8	9
0.50	−0.69315	.69115	.68916	.68717	.68518	.68320	.68122	.67924	.67727	.67531
.51	.67334	.67139	.66934	.66748	.66553	.66359	.66165	.65971	.65778	.65585
.52	.65393	.65201	.65009	.64817	.64626	.64436	.64245	.64055	.63866	.63677
.53	.63488	.63299	.63111	.62923	.62736	.62549	.62362	.62176	.61990	.61804
.54	.61619	.61434	.61249	.61065	.60881	.60697	.60514	.60331	.60148	.59966
.55	.59784	.59602	.59421	.59240	.59059	.58879	.58699	.58519	.58340	.58161
.56	.57982	.57803	.57625	.57448	.57270	.57093	.56916	.56740	.56563	.56387
.57	.56212	.56037	.55862	.55687	.55513	.55339	.55165	.54991	.54818	.54645
.58	.54473	.54300	.54128	.53957	.53785	.53614	.53444	.53273	.53103	.52933
.59	.52763	.52594	.52425	.52256	.52088	.51919	.51751	.51584	.51416	.51249
0.60	−0.51083	.50916	.50750	.50584	.50418	.50253	.50088	.49923	.49758	.49594
.61	.49430	.49266	.49102	.48939	.48776	.48613	.48451	.48289	.48127	.47965
.62	.47804	.47642	.47482	.47321	.47160	.47000	.46840	.46681	.46522	.46362
.63	.46204	.46045	.45887	.45728	.45571	.45413	.45256	.45099	.44942	.44785
.64	.44629	.44473	.44317	.44161	.44006	.43850	.43696	.43541	.43386	.43232
.65	.43078	.42925	.42771	.42618	.42465	.42312	.42159	.42007	.41855	.41703
.66	.41552	.41400	.41249	.41098	.40947	.40797	.40647	.40497	.40347	.40197
.67	.40048	.39899	.39750	.39601	.39453	.39304	.39156	.39008	.38861	.38713
.68	.38566	.38419	.38273	.38126	.37980	.37834	.37688	.37542	.37397	.37251
.69	.37106	.36962	.36817	.36673	.36528	.36384	.36241	.36097	.35954	.35810
0.70	−0.35667	.35525	.35382	.35240	.35098	.34956	.34814	.34672	.34531	.34390
.71	.34249	.34108	.33968	.33827	.33687	.33547	.33408	.33268	.33129	.32989
.72	.32850	.32712	.32573	.32435	.32296	.32158	.32021	.31883	.31745	.31608
.73	.31471	.31334	.31197	.31061	.30925	.30788	.30653	.30517	.30381	.30246
.74	.30111	.29975	.29841	.29706	.29571	.29437	.29303	.29169	.29035	.28902
.75	.28768	.28635	.28502	.28369	.28236	.28104	.27971	.27839	.27707	.27575
.76	.27444	.27312	.27181	.27050	.26919	.26788	.26657	.26527	.26397	.26266
.77	.26136	.26007	.25877	.25748	.25618	.25489	.25360	.25231	.25103	.24974
.78	.24846	.24718	.24590	.24462	.24335	.24207	.24080	.23953	.23826	.23699
.79	.23572	.23446	.23319	.23193	.23067	.22941	.22816	.22690	.22565	.22439
0.80	−0.22314	.22189	.22065	.21940	.21816	.21691	.21567	.21433	.21319	.21196
.81	.21072	.20949	.20825	.20702	.20579	.20457	.20334	.20212	.20089	.19967
.82	.19845	.19723	.19601	.19480	.19358	.19237	.19116	.18995	.18874	.18754
.83	.18633	.18513	.18392	.18272	.18152	.18032	.17913	.17793	.17674	.17554
.84	.17435	.17316	.17198	.17079	.16960	.16842	.16724	.16605	.16487	.16370
.85	−0.16252	.16134	.16017	.15900	.15782	.15665	.15548	.15432	.15315	.15199
.86	.15032	.14966	.14850	.14734	.14618	.14503	.14387	.14272	.14156	.14041
.87	.13926	.13811	.13697	.13582	.13467	.13353	.13239	.13125	.13011	.12897
.88	.12783	.12670	.12556	.12443	.12330	.12217	.12104	.11991	.11878	.11766
.89	.11653	.11541	.11429	.11317	.11205	.11093	.10981	.10870	.10759	.10647
0.90	−0.10536	.10425	.10314	.10203	.10093	.09982	.09872	.09761	.09651	.09541
.91	.09431	.09321	.09212	.09102	.08992	.08883	.08744	.08665	.08556	.08447
.92	.08338	.08230	.08121	.08013	.07904	.07796	.07688	.07580	.07472	.07365
.93	.07257	.07150	.07042	.06935	.06828	.06721	.06614	.06507	.06401	.06294
.94	.06188	.06081	.05975	.05869	.05763	.05657	.05551	.05446	.05340	.05235
.95	.05129	.05024	.04919	.04814	.04709	.04604	.04500	.04395	.04291	.04186
.96	.04082	.03978	.03874	.03770	.03666	.03563	.03459	.03356	.03252	.03149
.97	.03046	.02943	.02840	.02737	.02634	.02532	.02429	.02327	.02225	.02122
.98	.02020	.01918	.01816	.01715	.01613	.01511	.01410	.01309	.01207	.01106
.99	.01005	.00904	.00803	.00702	.00602	.00501	.00401	.00300	.00200	.00100

TABLE IV. NATURAL OR NAPERIAN LOGARITHMS (Continued)
0–499

N	0	1	2	3	4	5	6	7	8	9
0	− ∞	0.00000	0.69315	1.09861	.38629	60944	.79176	.94591	*.07944	*.19722
1	2.30259	.39790	.48491	.56495	.63906	.70805	.77259	.83321	.89037	.94444
2	.99573	*.04452	*.09104	*.13549	*.17805	*.21888	*.25810	*.29584	*.33220	*.36730
3	3.40120	.43399	.46574	.49651	.52636	.55535	.58352	.61092	.63759	.66356
4	.68888	.71357	.73767	.76120	.78419	.80666	.82864	.85015	.87120	.89182
5	.91202	.93183	.95124	.97029	.98898	*.00733	*.02535	*.04305	*.06044	*.07754
6	4.09434	.11087	.12713	.14313	.15888	.17439	.18965	.20469	.21951	.23411
7	.24850	.26268	.27667	.29046	.30407	.31749	.33073	.34381	.35671	.36945
8	.38203	.39445	.40672	.41884	.43082	.44265	.45435	.46591	.47734	.48864
9	.49981	.51086	.52179	.53260	.54329	.55388	.56435	.57471	.58497	.59512
10	4.60517	.61512	.62497	.63473	.64439	.65396	.66344	.67283	.68213	.69135
11	.70048	.70953	.71850	.72739	.73620	.74493	.75359	.76217	.77068	.77912
12	.78749	.79579	.80402	.81218	.82028	.82831	.83628	.84419	.85203	.85981
13	.86753	.87520	.88280	.89035	.89784	.90527	.91265	.91998	.92725	.93447
14	.94164	.94876	.95583	.96284	.96981	.97673	.98361	.99043	.99721	*.00395
15	5.01064	.01728	.02388	.03044	.03695	.04343	.04986	.05625	.06260	.06890
16	.07517	.08140	.08760	.09375	.09987	.10595	.11199	.11799	.12396	.12990
17	.13580	.14166	.14749	.15329	.15906	.16479	.17048	.17615	.18178	.18739
18	.19296	.19850	.20401	.20949	.21494	.22036	.22575	.23111	.23644	.24175
19	.24702	.25227	.25750	.26269	.26786	.27300	.27811	.28320	.28827	.29330
20	5.29832	.30330	.30827	.31321	.31812	.32301	.32788	.33272	.33754	.34233
21	.34711	.35186	.35659	.36129	.36598	.37064	.37528	.37990	.38450	.38907
22	.39363	.39816	.40268	.40717	.41165	.41610	.42053	.42495	.42935	.43372
23	.43808	.44242	.44674	.45104	.45532	.45959	.46383	.46806	.47227	.47646
24	.48064	.48480	.48894	.49306	.49717	.50126	.50533	.50939	.51343	.51745
25	.52146	.52545	.52943	.53339	.53733	.54126	.54518	.54908	.55296	.55683
26	.56068	.56452	.56834	.57215	.57595	.57973	.58350	.58725	.59099	.59471
27	.59842	.60212	.60580	.60947	.61313	.61677	.62040	.62402	.62762	.63121
28	.63479	.63835	.64191	.64545	.64897	.65249	.65599	.65948	.66296	.66643
29	.66988	.67332	.67675	.68017	.68358	.68698	.69036	.69373	.69709	.70044
30	5.70378	.70711	.71043	.71373	.71703	.72031	.72359	.72685	.73010	.73334
31	.73657	.73979	.74300	.74620	.74939	.75257	.75574	.75890	.76205	.76519
32	.76832	.77144	.77455	.77765	.78074	.78383	.78690	.78996	.79301	.79606
33	.79909	.80212	.80513	.80814	.81114	.81413	.81711	.82008	.82305	.82600
34	.82895	.83188	.83481	.83773	.84064	.84354	.84644	.84932	.85220	.85507
35	.85793	.86079	.86363	.86647	.86930	.87212	.87493	.87774	.88053	.88332
36	.88610	.88888	.89164	.89440	.89715	.89990	.90263	.90536	.90808	.91080
37	.91350	.91620	.91889	.92158	.92426	.92693	.92959	.93225	.93489	.93754
38	.94017	.94280	.94542	.94803	.95064	.95324	.95584	.95842	.96101	.96358
39	.96615	.96871	.97126	.97381	.97635	.97889	.98141	.98394	.98645	.98896
40	5.99146	.99396	.99645	.99894	*.00141	*.00389	*.00635	*.00881	*.01127	*.01372
41	6.01616	.01859	.02102	.02345	.02587	.02828	.03069	.03309	.03548	.03787
42	.04025	.04263	.04501	.04737	.04973	.05209	.05444	.05678	.05192	.06146
43	.06379	.06611	.06843	.07074	.07304	.07535	.07764	.07993	.08222	.08450
44	.08677	.08904	.09131	.09357	.09582	.09807	.10032	.10256	.10479	.10702
45	.10925	.11147	.11368	.11589	.11810	.12030	.12249	.12468	.12687	.12905
46	.13123	.13340	.13556	.13773	.13988	.14204	.14419	.14633	.14847	.15060
47	.15273	.15486	.15698	.15910	.16121	.16331	.16542	.16752	.16961	.17170
48	.17379	.17587	.17794	.18002	.18208	.18415	.18621	.18826	.19032	.19236
49	.19441	.19644	.19848	.20051	.20254	.20456	.20658	.20859	.21060	.21261

TABLE IV. NATURAL OR NAPERIAN LOGARITHMS (Continued)
500–999

N	0	1	2	3	4	5	6	7	8	9
50	6.21461	.21661	.21860	.22059	.22258	.22456	.22654	.22851	.23048	.23245
51	.23441	.23637	.23832	.24028	.24222	.24417	.24611	.24804	.24998	.25190
52	.25383	.25575	.25767	.25958	.26149	.26340	.26530	.26720	.26910	.27099
53	.27288	.27476	.27664	.27852	.28040	.28227	.28413	.28600	.28786	.28972
54	.29157	.29342	.29527	.29711	.29895	.30079	.30262	.30445	.30628	.30810
55	.30992	.31173	.31355	.31536	.31716	.31897	.32077	.32257	.32436	.32615
56	.32794	.32972	.33150	.33328	.33505	.33683	.33859	.34036	.34212	.34388
57	.34564	.34739	.34914	.35089	.35263	.35437	.35611	.35784	.35957	.36130
58	.36303	.36475	.36647	.36819	.36990	.37161	.37332	.37502	.37673	.37843
59	.38012	.38182	.38351	.38519	.38688	.38856	.39024	.39192	.39359	.39526
60	6.30693	.39859	.40026	.40192	.40357	.40523	.40688	.40853	.41017	.41182
61	.41346	.41510	.41673	.41836	.41999	.42162	.42325	.42487	.42649	.42811
62	.42972	.43133	.43294	.43455	.43615	.43775	.43935	.44095	.44254	.44413
63	.44572	.44731	.44889	.45047	.45205	.45362	.45520	.45677	.45834	.45990
64	.46147	.46303	.46459	.46614	.46770	.46925	.47080	.47235	.47389	.47543
65	.47697	.47851	.48004	.48158	.48311	.48464	.48616	.48768	.48920	.49072
66	.49224	.49375	.49527	.49677	.49828	.49979	.50129	.50279	.50429	.50578
67	.50728	.50877	.51026	.51175	.51323	.51471	.51619	.51767	.51915	.52062
68	.52209	.52356	.52503	.52649	.52796	.52942	.53088	.53233	.53379	.53524
69	.53669	.53814	.53959	.54103	.54247	.54391	.54535	.54679	.54822	.54965
70	6.55108	.55251	.55393	.55536	.55678	.55820	.55962	.56103	.56244	.56386
71	.56526	.56667	.56808	.56948	.57088	.57228	.57368	.57508	.57647	.57786
72	.57925	.58064	.58203	.58341	.58479	.58617	.58755	.58893	.59030	.59167
73	.59304	.59441	.59578	.59715	.59851	.59987	.60123	.60259	.60394	.60530
74	.60665	.60800	.60935	.61070	.61204	.61338	.61473	.61607	.61740	.61874
75	.62007	.62141	.62274	.62407	.62539	.62672	.62804	.62936	.63068	.63200
76	.63332	.63463	.63595	.63726	.63857	.63988	.64118	.64249	.64379	.64509
77	.64639	.64769	.64898	.65028	.65157	.65286	.65415	.65544	.65673	.65801
78	.65929	.66058	.66185	.66313	.66441	.66568	.66696	.66823	.66950	.67077
79	.67203	.67330	.67456	.67582	.67708	.67834	.67960	.68085	.68211	.68336
80	6.68461	.68586	.68711	.68835	.68960	.69084	.69208	.69332	.69456	.69580
81	.69703	.69827	.69950	.70073	.70196	.70319	.70441	.70564	.70686	.70808
82	.70930	.71052	.71174	.71296	.71417	.71538	.71659	.71780	.71901	.72022
83	.72143	.72263	.72383	.72503	.72623	.72743	.72863	.72982	.73102	.73221
84	.73340	.73459	.73578	.73697	.73815	.73934	.74052	.74170	.74288	.74406
85	.74524	.74641	.74759	.74876	.74993	.75110	.75227	.75344	.75460	.75577
86	.75693	.75809	.75926	.76041	.76157	.76273	.76388	.76504	.76619	.76734
87	.76849	.76964	.77079	.77194	.77308	.77422	.77537	.77651	.77765	.77878
88	.77992	.78106	.78219	.78333	.78446	.78559	.78672	.78784	.78897	.79010
89	.79122	.79234	.79347	.79459	.79571	.79682	.79794	.79906	.80017	.80128
90	6.80239	.80351	.80461	.80572	.80683	.80793	.80904	.81014	.81124	.81235
91	.81344	.81454	.81564	.81674	.81783	.81892	.82002	.82111	.82220	.82329
92	.82437	.82546	.82655	.82763	.82871	.82979	.83087	.83195	.83303	.83411
93	.83518	.83626	.83733	.83841	.83948	.84055	.84162	.84268	.84375	.84482
94	.84588	.84694	.84801	.84907	.85013	.85118	.85224	.85330	.85435	.85541
95	.85646	.85751	.85857	.85961	.86066	.86171	.86276	.86380	.86485	.86589
96	.86693	.86797	.86901	.87005	.87109	.87213	.87316	.87420	.87523	.87626
97	.87730	.87833	.87936	.88038	.88141	.88244	.88346	.88449	.88551	.88653
98	.88755	.88857	.88959	.89061	.89163	.89264	.89366	.89467	.89568	.89669
99	.89770	.89871	.89972	.90073	.90174	.90274	.90375	.90475	.90575	.90675

TABLE V. EXPONENTIAL FUNCTIONS

x	e^x	e^{-x}	x	e^x	e^{-x}
0.0	1.000	1.000	**5.0**	148.4	0.0067
0.1	1.105	0.905	**5.1**	164.0	0.0061
0.2	1.221	0.819	**5.2**	181.3	0.0055
0.3	1.350	0.741	**5.3**	200.3	0.0050
0.4	1.492	0.670	**5.4**	221.4	0.0045
0.5	1.649	0.607	**5.5**	244.7	0.0041
0.6	1.822	0.549	**5.6**	270.4	0.0037
0.7	2.014	0.497	**5.7**	298.9	0.0033
0.8	2.226	0.449	**5.8**	330.3	0.0030
0.9	2.460	0.407	**5.9**	365.0	0.0027
1.0	2.718	0.368	**6.0**	403.4	0.0025
1.1	3.004	0.333	**6.1**	445.9	0.0022
1.2	3.320	0.301	**6.2**	492.8	0.0020
1.3	3.669	0.273	**6.3**	544.6	0.0018
1.4	4.055	0.247	**6.4**	601.8	0.0017
1.5	4.482	0.223	**6.5**	665.1	0.0015
1.6	4.953	0.202	**6.6**	735.1	0.0014
1.7	5.474	0.183	**6.7**	812.4	0.0012
1.8	6.050	0.165	**6.8**	897.8	0.0011
1.9	6.686	0.150	**6.9**	992.3	0.0010
2.0	7.389	0.135	**7.0**	1,096.6	0.0009
2.1	8.166	0.122	**7.1**	1,212.0	0.0008
2.2	9.025	0.111	**7.2**	1,339.4	0.0007
2.3	9.974	0.100	**7.3**	1,480.3	0.0007
2.4	11.023	0.091	**7.4**	1,636.0	0.0006
2.5	12.18	0.082	**7.5**	1,808.0	0.00055
2.6	13.46	0.074	**7.6**	1,998.2	0.00050
2.7	14.88	0.067	**7.7**	2,208.3	0.00045
2.8	16.44	0.061	**7.8**	2,440.6	0.00041
2.9	18.17	0.055	**7.9**	2,697.3	0.00037
3.0	20.09	0.050	**8.0**	2,981.0	0.00034
3.1	22.20	0.045	**8.1**	3,294.5	0.00030
3.2	24.53	0.041	**8.2**	3,641.0	0.00027
3.3	27.11	0.037	**8.3**	4,023.9	0.00025
3.4	29.96	0.033	**8.4**	4,447.1	0.00022
3.5	33.12	0.030	**8.5**	4,914.8	0.00020
3.6	36.60	0.027	**8.6**	5,431.7	0.00018
3.7	40.45	0.025	**8.7**	6,002.9	0.00017
3.8	44.70	0.022	**8.8**	6,634.2	0.00015
3.9	49.40	0.020	**8.9**	7,332.0	0.00014
4.0	54.60	0.018	**9.0**	8,103.1	0.00012
4.1	60.34	0.017	**9.1**	8,955.3	0.00011
4.2	66.69	0.015	**9.2**	9,897.1	0.00010
4.3	73.70	0.014	**9.3**	10,938	0.00009
4.4	81.45	0.012	**9.4**	12,088	0.00008
4.5	90.02	0.011	**9.5**	13,360	0.00007
4.6	99.48	0.010	**9.6**	14,765	0.00007
4.7	109.95	0.009	**9.7**	16,318	0.00006
4.8	121.51	0.008	**9.8**	18,034	0.00006
4.9	134.29	0.007	**9.9**	19,930	0.00005

TABLE VI. TABLE OF INTEGRALS

Elementary Forms

1. $\int [f(x) + g(x) - h(x)]\,dx = \int f(x)\,dx + \int g(x)\,dx - \int h(x)\,dx$

2. $\int kf(x)\,dx = k\int f(x)\,dx$

3. $\int dx = x + c$

4. $\int k\,dx = kx + c$

5. $\int x^n\,dx = \dfrac{x^{n+1}}{n+1} + c \qquad \text{if } n \neq -1$

6. $\int x^{-1}\,dx = \ln x + c$

Forms Containing $a + bx$

7. $\int (a + bx)^n\,dx = \dfrac{1}{b(n+1)}(a + bx)^{n+1} + c \qquad \text{if } n \neq -1$

$\qquad = \dfrac{1}{b}\ln(a + bx) + c \qquad \text{if } n = -1$

8. $\int x(a + bx)^n\,dx = \dfrac{(a + bx)^{n+2}}{b^2(n+2)} - \dfrac{(a + bx)^{n+1}}{b^2(n+1)} + c \qquad \text{if } n \neq -1, -2$

$\qquad = \dfrac{1}{b^2}[a + bx - a\ln(a + bx)] + c \qquad \text{if } n = -1$

$\qquad = \dfrac{1}{b^2}\left[\ln(a + bx) + \dfrac{a}{a + bx}\right] + c \qquad \text{if } n = -2$

Forms Containing $x^2 \pm a^2$, or $a^2 \pm x^2$

9. $\int \dfrac{dx}{a^2 - x^2} = \dfrac{1}{2a}\ln\left(\dfrac{a + x}{a - x}\right) + c$

10. $\int \dfrac{dx}{x^2 - a^2} = \dfrac{1}{2a}\ln\left(\dfrac{x - a}{x + a}\right) + c$

11. $\int (x^2 \pm a^2)^{1/2}\,dx = \frac{1}{2}[x(x^2 \pm a^2)^{1/2} \pm a^2\ln(x + (x^2 \pm a^2)^{1/2}] + c$

12. $\int (x^2 \pm a^2)^{-1/2}\,dx = \ln[x + (x^2 \pm a^2)^{1/2}] + c$

13. $\int x(x^2 \pm a^2)^{1/2}\,dx = \frac{1}{3}(x^2 \pm a^2)^{3/2} + c$

14. $\int x(x^2 \pm a^2)^{-1/2}\, dx = (x^2 \pm a^2)^{1/2} + c$

15. $\int x(a^2 - x^2)^{1/2}\, dx = -\frac{1}{3}(a^2 - x^2)^{3/2} + c$

16. $\int x(a^2 - x^2)^{-1/2}\, dx = -(a^2 - x^2)^{1/2} + c$

Exponential Forms

17. $\int e^x\, dx = e^x + c$

18. $\int a^{kx}\, dx = \frac{a^{kx}}{k \ln a} + c$

19. $\int xe^{kx}\, dx = \frac{e^{kx}}{k^2}(kx - 1) + c$

20. $\int e^{kx}\, dx = \frac{e^{kx}}{k} + c$

Logarithmic Forms

21. $\int \ln x\, dx = x \ln x - x + c$

22. $\int \log_a x\, dx = x \log_a x - x \log_a e + c$

23. $\int \frac{dx}{x (\ln x)} = \ln (\ln x) + c$

24. $\int (\ln x)^2\, dx = x(\ln x)^2 - 2x(\ln x) + 2x + c$

25. $\int x \ln x\, dx = \frac{x^2 \ln x}{2} - \frac{x^2}{4} + c$

TABLE VII. SQUARES AND SQUARE ROOTS
1–1000

N	N^2	$\sqrt{N}$	N	N^2	$\sqrt{N}$	N	N^2	$\sqrt{N}$
			50	2 500	7.071 068	100	10 000	10.00000
1	1	1.000 000	51	2 601	7.141 428	101	10 201	10.04988
2	4	1.414 214	52	2 704	7.211 103	102	10 404	10.09950
3	9	1.732 051	53	2 809	7.280 110	103	10 609	10.14889
4	16	2.000 000	54	2 916	7.348 469	104	10 816	10.19804
5	25	2.236 068	55	3 025	7.416 198	105	11 025	10.24695
6	36	2.449 490	56	3 136	7.483 315	106	11 236	10.29563
7	49	2.645 751	57	3 249	7.549 834	107	11 449	10.34408
8	64	2.828 427	58	3 364	7.615 773	108	11 664	10.39230
9	81	3.000 000	59	3 481	7.681 146	109	11 881	10.44031
10	100	3.162 278	60	3 600	7.745 967	110	12 100	10.48809
11	121	3.316 625	61	3 721	7.810 250	111	12 321	10.53565
12	144	3.464 102	62	3 844	7.874 008	112	12 544	10.58301
13	169	3.605 551	63	3 969	7.937 254	113	12 769	10.63015
14	196	3.741 657	64	4 096	8.000 000	114	12 906	10.67708
15	225	3.872 983	65	4 225	8.062 258	115	13 225	10.72381
16	256	4.000 000	66	4 356	8.124 038	116	13 456	10.77033
17	289	4.123 106	67	4 489	8.185 353	117	13 689	10.81665
18	324	4.242 641	68	4 624	8.246 211	118	13 924	10.86278
19	361	4.358 899	69	4 761	8.306 624	119	14 161	10.90871
20	400	4.472 136	70	4 900	8.366 600	120	14 400	10.95445
21	441	4.582 576	71	5 041	8.426 150	121	14 641	11.00000
22	484	4.690 416	72	5 184	8.485 281	122	14 884	11.04536
23	529	4.795 832	73	5 329	8.544 004	123	15 129	11.09054
24	576	4.898 979	74	5 476	8.602 325	124	15 376	11.13553
25	625	5.000 000	75	5 625	8.660 254	125	15 625	11.18034
26	676	5.099 020	76	5 776	8.717 798	126	15 876	11.22497
27	729	5.196 152	77	5 929	8.774 964	127	16 129	11.26943
28	784	5.291 503	78	6 084	8.831 761	128	16 384	11.31371
29	841	5.385 165	79	6 241	8.888 194	129	16 641	11.35782
30	900	5.477 226	80	6 400	8.944 272	130	16 900	11.40175
31	961	5.567 764	81	6 561	9.000 000	131	17 161	11.44552
32	1 024	5.656 854	82	6 724	9.055 385	132	17 424	11.48913
33	1 089	5.744 563	83	6 889	9.110 434	133	17 689	11.53256
34	1 156	5.830 952	84	7 056	9.165 151	134	17 956	11.57584
35	1 225	5.916 080	85	7 225	9.219 544	135	18 225	11.61895
36	1 296	6.000 000	86	7 396	9.273 618	136	18 496	11.66190
37	1 369	6.082 763	87	7 569	9.327 379	137	18 769	11.70470
38	1 444	6.164 414	88	7 744	9.380 832	138	19 044	11.74734
39	1 521	6.244 998	89	7 921	9.433 981	139	19 321	11.78983
40	1 600	6.324 555	90	8 100	9.486 833	140	19 600	11.83216
41	1 681	6.403 124	91	8 281	9.539 392	141	19 881	11.87434
42	1 764	6.480 741	92	8 464	9.591 663	142	20 164	11.91638
43	1 849	6.557 439	93	8 649	9.643 651	143	20 449	11.95826
44	1 936	6.633 250	94	8 836	9.695 360	144	20 736	12.00000
45	2 025	6.708 204	95	9 025	9.746 794	145	21 025	12.04159
46	2 116	6.782 330	96	9 216	9.797 959	146	21 316	12.08305
47	2 209	6.855 655	97	9 409	9.848 858	147	21 609	12.12436
48	2 304	6.928 203	98	9 604	9.899 405	148	21 904	12.16553
49	2 401	7.000 000	99	9 801	9.949 874	149	22 201	12.20656

TABLE VII. SQUARES AND SQUARE ROOTS (Continued)
1–1000

N	N^2	$\sqrt{N}$	N	N^2	$\sqrt{N}$	N	N^2	$\sqrt{N}$
150	22 500	12.24745	200	40 000	14.14214	250	62 500	15.81139
151	22 801	12.28821	201	40 401	14.17745	251	63 001	15.84298
152	23 104	12.32883	202	40 804	14.21267	252	63 504	15.87451
153	23 409	12.36932	203	41 209	14.24781	253	64 009	15.90597
154	23 716	12.40967	204	41 616	14.28286	254	64 516	15.93738
155	24 025	12.44990	205	42 025	14.31782	255	65 025	15.96872
156	24 336	12.49000	206	42 436	14.35270	256	65 536	16.00000
157	24 649	12.52996	207	42 849	14.38749	257	66 049	16.03122
158	24 964	12.56981	208	43 264	14.42221	258	66 564	16.06238
159	25 281	12.60952	209	43 681	14.45683	259	67 081	16.09348
160	25 600	12.64911	210	44 100	14.49138	260	67 600	16.12452
161	25 921	12.68858	211	44 521	14.52584	261	68 121	16.15549
162	26 244	12.72792	212	44 944	14.56022	262	68 644	16.18641
163	26 569	12.76715	213	45 369	14.59452	263	69 169	16.21727
164	26 896	12.80625	214	45 796	14.62874	264	69 696	16.24808
165	27 225	12.84523	215	46 225	14.66288	265	70 225	16.27882
166	27 556	12.88410	216	46 656	14.69694	266	70 756	16.30951
167	27 889	12.92285	217	47 089	14.73092	267	71 289	16.34013
168	28 224	12.96148	218	47 524	14.76482	268	71 824	16.37071
169	28 561	13.00000	219	47 961	14.79865	269	72 361	16.40122
170	28 900	13.03840	220	48 400	14.83240	270	72 900	16.43168
171	29 241	13.07670	221	48 841	14.86607	271	73 441	16.46208
172	29 584	13.11488	222	49 284	14.89966	272	73 984	16.49242
173	29 929	13.15295	223	49 729	14.93318	273	74 529	16.52271
174	30 276	13.19091	224	50 176	14.96663	274	75 076	16.55295
175	30 625	13.22876	225	50 625	15.00000	275	75 625	16.58312
176	30 976	13.26650	226	51 076	15.03330	276	76 176	16.61325
177	31 329	13.30413	227	51 529	15.06652	277	76 729	16.64332
178	31 684	13.34166	228	51 984	15.09967	278	77 284	16.67333
179	32 041	13.37909	229	52 441	15.13275	279	77 841	16.70329
180	32 400	13.41641	230	52 900	15.16575	280	78 400	16.73320
181	32 761	13.45362	231	53 361	15.19868	281	78 961	16.76305
182	33 124	13.49074	232	53 824	15.23155	282	79 524	16.79286
183	33 489	13.52775	233	54 289	15.26434	283	80 089	16.82260
184	33 856	13.56466	234	54 756	15.29706	284	80 656	16.85230
185	34 225	13.60147	235	55 225	15.32971	285	81 225	16.88194
186	34 596	13.63818	236	55 696	15.36229	286	81 796	16.91153
187	34 969	13.67479	237	56 169	15.39480	287	82 369	16.94107
188	35 344	13.71131	238	56 644	15.42725	288	82 944	16.97056
189	35 721	13.74773	239	57 121	15.45962	289	83 521	17.00000
190	36 100	13.78405	240	57 600	15.49193	290	84 100	17.02939
191	36 481	13.82027	241	58 081	15.52417	291	84 681	17.05872
192	36 864	13.85641	242	58 564	15.55635	292	85 264	17.08801
193	37 249	13.89244	243	59 049	15.58846	293	85 849	17.11724
194	37 636	13.92839	244	59 536	15.62050	294	86 436	17.14643
195	38 025	13.96424	245	60 025	15.65248	295	87 025	17.17556
196	38 416	14.00000	246	60 516	15.68439	296	87 616	17.20465
197	38 809	14.03567	247	61 009	15.71623	297	88 209	17.23369
198	39 204	14.07125	248	61 504	15.74802	298	88 804	17.26268
199	39 601	14.10674	249	62 001	15.77973	299	89 401	17.29162

TABLE VII. SQUARES AND SQUARE ROOTS (Continued)

1–1000

N	N^2	$\sqrt{N}$	N	N^2	$\sqrt{N}$	N	N^2	$\sqrt{N}$
300	90 000	17.32051	350	122 500	18.70829	400	160 000	20.00000
301	90 601	17.34935	351	123 201	18.73499	401	160 801	20.02498
302	91 204	17.37815	352	123 904	18.76166	402	161 604	20.04994
303	91 809	17.40690	353	124 609	18.78829	403	162 409	20.07486
304	92 416	17.43560	354	125 316	18.81489	404	163 216	20.09975
305	93 025	17.46425	355	126 025	18.84144	405	164 025	20.12461
306	93 636	17.49286	356	126 736	18.86796	406	164 836	20.14944
307	94 249	17.52142	357	127 449	18.89444	407	165 649	20.17424
308	94 864	17.54993	358	128 164	18.92089	408	166 464	20.19901
309	95 481	17.57840	359	128 881	18.94730	409	167 231	20.22375
310	96 100	17.60682	360	129 600	18.97367	410	168 100	20.24846
311	96 721	17.63519	361	130 321	19.00000	411	168 921	20.27313
312	97 344	17.66352	362	131 044	19.02630	412	169 744	20.29778
313	97 969	17.69181	363	131 769	19.05256	413	170 569	20.32249
314	98 596	17.72005	364	132 496	19.07878	414	171 396	20.34699
315	99 225	17.74824	365	133 225	19.10497	415	172 225	20.37155
316	99 856	17.77639	366	133 956	19.13113	416	173 056	20.39608
317	100 489	17.80449	367	134 689	19.15724	417	173 889	20.42058
318	101 124	17.83255	368	135 424	19.18333	418	174 724	20.44505
319	101 761	17.86057	369	136 161	19.20937	419	175 561	20.46949
320	102 400	17.88854	370	136 900	19.23538	420	176 400	20.49390
321	103 041	17.91647	371	137 641	19.26136	421	177 241	20.51828
322	103 684	17.94436	372	138 384	19.28730	422	178 084	20.54264
323	104 329	17.97220	373	139 129	19.31321	423	178 929	20.56696
324	104 976	18.00000	374	139 876	19.33908	424	179 776	20.59126
325	105 625	18.02776	375	140 625	19.36492	425	180 625	20.61553
326	106 276	18.05547	376	141 376	19.39072	426	181 476	20.63977
327	106 929	18.08314	377	142 129	19.41649	427	182 329	20.66398
328	107 584	18.11077	378	142 884	19.44222	428	183 184	20.68816
329	108 241	18.13836	379	143 641	19.46792	429	184 041	20.71232
330	108 900	18.16590	380	144 400	19.49359	430	184 900	20.73644
331	109 561	18.19341	381	145 161	19.51922	431	185 761	20.76054
332	110 224	18.22087	382	145 924	19.54483	432	186 624	20.78461
333	110 889	18.24829	383	146 689	19.57039	433	187 489	20.80865
334	111 556	18.27567	384	147 456	19.59592	434	188 356	20.83267
335	112 225	18.30301	385	148 225	19.62142	435	189 225	20.85665
336	112 896	18.33030	386	148 996	19.64688	436	190 096	20.88061
337	113 569	18.35756	387	149 769	19.67232	437	190 969	20.90454
338	114 244	18.38478	388	150 544	19.69772	438	191 844	20.92845
339	114 921	18.41195	389	151 321	19.72308	439	192 721	20.95233
340	115 600	18.43909	390	152 100	19.74842	440	193 600	20.97618
341	116 281	18.46619	391	152 881	19.77372	441	194 481	21.00000
342	116 964	18.49324	392	153 664	19.79899	442	195 364	21.02380
343	117 649	18.52026	393	154 449	19.82423	443	196 249	21.04757
344	118 336	18.54724	394	155 236	19.84943	444	197 136	21.07131
345	119 025	18.57418	395	156 025	19.87461	445	198 025	21.09502
346	119 716	18.60108	396	156 816	19.89975	446	198 916	21.11871
347	120 409	18.62794	397	157 609	19.92486	447	199 809	21.14237
348	121 104	18.65476	398	158 404	19.94994	448	200 704	21.16601
349	121 801	18.68154	399	159 201	19.97498	449	201 601	21.18962

TABLE VII. SQUARES AND SQUARE ROOTS (Continued)
1–1000

N	N^2	$\sqrt{N}$	N	N^2	$\sqrt{N}$	N	N^2	$\sqrt{N}$
450	202 500	21.21320	500	250 000	22.36068	550	302 500	23.45208
451	203 401	21.23676	501	251 001	22.38303	551	303 601	23.47339
452	204 304	21.26029	502	252 004	22.40536	552	304 704	23.49468
453	205 209	21.28380	503	253 009	22.42766	553	305 809	23.51595
454	206 116	21.30728	504	254 016	22.44994	554	306 916	23.53720
455	207 025	21.33073	505	255 025	22.47221	555	308 025	23.55844
456	207 936	21.35416	506	256 036	22.49444	556	309 136	23.57965
457	208 849	21.37756	507	257 049	22.51666	557	310 249	23.60085
458	209 764	21.40093	508	258 064	22.53886	558	311 364	23.62202
459	210 681	21.42429	509	259 081	22.56103	559	312 481	23.64318
460	211 600	21.44761	510	260 100	22.58318	560	313 600	23.66432
461	212 521	21.47091	511	261 121	22.60531	561	314 721	23.68544
462	213 444	21.49419	512	262 144	22.62742	562	315 844	23.70654
463	214 369	21.51743	513	263 169	22.64950	563	316 969	23.72762
464	215 296	21.54066	514	264 196	22.67157	564	318 096	23.74868
465	216 225	21.56386	515	265 225	22.69361	565	319 225	23.76973
466	217 156	21.58703	516	266 256	22.71563	566	320 356	23.79075
467	218 089	21.61018	517	267 289	22.73763	567	321 489	23.81176
468	219 024	21.63331	518	268 324	22.75961	568	322 624	23.83275
469	219 961	21.65641	519	269 361	22.78157	569	323 761	23.85372
470	220 900	21.67948	520	270 400	22.80351	570	324 900	23.87467
471	221 841	21.70253	521	271 441	22.82542	571	326 041	23.89561
472	222 781	21.72556	522	272 484	22.84732	572	327 184	23.91652
473	223 729	21.74856	523	273 529	22.86919	573	328 329	23.93742
474	224 676	21.77154	524	274 576	22.89105	574	329 476	23.95830
475	225 625	21.79449	525	275 625	22.91288	575	330 625	23.97916
476	226 576	21.81742	526	276 676	22.93469	576	331 776	24.00000
477	227 529	21,84033	527	277 729	22.95648	577	332 929	24.02082
478	228 484	21.86321	528	278 784	22.97825	578	334 084	24.04163
479	229 441	21.88607	529	279 841	23.00000	579	335 241	24.06242
480	230 400	21.90890	530	280 900	23.02173	580	336 400	24.08319
481	231 361	21.93171	531	281 961	23.04344	581	337 561	24.10394
482	232 324	21.95450	532	283 024	23.06513	582	338 724	24.12468
483	233 289	21.97726	533	284 089	23.08679	583	339 889	24.14539
484	234 256	22.00000	534	285 156	23.10844	584	341 056	24.16609
485	235 225	22.02272	535	286 225	23.13007	585	342 225	24.18677
486	236 196	22.04541	536	287 296	23.15167	586	343 396	24.20744
487	237 169	22.06808	537	288 369	23.17326	587	344 569	24.22808
488	238 144	22.09072	538	289 444	23.19483	588	345 744	24.24871
489	239 121	22.11334	539	290 521	23.21637	589	346 921	24.26932
490	240 100	22.13594	540	291 600	23.23790	590	348 100	24.28992
491	241 081	22.15852	541	292 681	23.25941	591	349 281	24.31049
492	242 064	22.18107	542	293 764	23.28089	592	350 464	24.33105
493	243 049	22.20360	543	294 849	23.30236	593	351 649	24.35159
494	244 036	22.22611	544	295 936	23.32381	594	352 836	24.37212
495	245 025	22.24860	545	297 025	23.34524	595	354 025	24.39262
496	246 016	22.27106	546	298 116	23.36664	596	355 216	24.41311
497	247 009	22.29350	547	299 209	23.38803	597	356 409	24.43358
498	248 004	22.31591	548	300 304	23.40940	598	357 604	24.45404
499	249 001	22.33831	549	301 401	23.43075	599	358 801	24.47448

TABLE VII. SQUARES AND SQUARE ROOTS (Continued)

1–1000

N	N^2	$\sqrt{N}$	N	N^2	$\sqrt{N}$	N	N^2	$\sqrt{N}$
600	360 000	24.49490	650	422 500	25.49510	700	490 000	26.45751
601	361 201	24.51530	651	423 801	25.51470	701	491 401	26.47640
602	362 404	24.53569	652	425 104	25.53429	702	492 804	26.49528
603	363 609	24.55606	653	426 409	25.56386	703	494 209	26.51415
604	364 816	24.57641	654	427 716	25.57342	704	495 616	26.53200
605	366 025	24.59675	655	429 025	25.59297	705	497 025	26.55184
606	367 236	24.61707	656	430 336	25.61250	706	498 436	26.57066
607	368 449	24.63737	657	431 649	25.63201	707	499 849	26.58947
608	369 664	24.65766	658	432 964	25.65151	708	501 264	26.60827
609	370 881	24.67793	659	434 281	25.67100	709	502 681	26.62705
610	372 100	24.69818	660	435 600	25.69047	710	504 100	26.64583
611	373 321	24.71841	661	436 921	25.70992	711	505 521	26.66458
612	374 544	24.73863	662	438 244	25.72936	712	506 944	26.68383
613	375 769	24.75884	663	439 569	25.74829	713	508 369	26.70206
614	376 996	24.77902	664	440 896	25.76820	714	509 796	26.72078
615	378 225	24.79919	665	442 225	25.78759	715	511 225	26.73948
616	379 456	24.81935	666	443 556	25.80698	716	512 656	26.75818
617	380 689	24.83948	667	444 889	25.82634	717	514 089	26.77686
618	381 924	24.85961	668	446 224	25.84570	718	515 524	26.79552
619	383 161	24.87971	669	447 561	25.86503	719	516 961	26.81418
620	384 400	24.89980	670	448 900	25.88436	720	518 400	26.83282
621	385 641	24.91987	671	450 241	25.90367	721	519 841	26.85144
622	386 884	24.93993	672	451 584	25.92296	722	521 284	26.87006
623	388 129	24.95997	673	452 929	25.94224	723	522 729	26.88966
624	389 376	24.97999	674	454 276	25.96151	724	524 176	26.90725
625	390 625	25.00000	675	455 625	25.98076	725	525 625	26.92582
626	391 876	25.01999	676	456 976	26.00000	726	527 076	26.94439
627	393 129	25.03997	677	458 329	26.01922	727	528 529	26.96294
628	394 384	25.05993	678	459 684	26.03843	728	529 984	26.98148
629	395 641	25.07987	679	461 041	26.05763	729	531 441	27.00000
630	396 900	25.09980	680	462 400	26.07681	730	532 900	27.01851
631	398 161	25.11971	681	463 761	26.09598	731	534 361	27.03701
632	399 424	25.13961	682	465 124	26.11513	732	535 824	27.03550
633	400 689	25.15949	683	466 489	26.13427	733	537 289	27.02397
634	401 956	25.17936	684	467 856	26.15339	734	538 756	27.09243
635	403 225	25.19921	685	469 225	26.17250	735	540 225	27.11088
636	404 496	25.21904	686	470 596	26.19160	736	541 696	27.12932
637	405 769	25.23886	687	471 969	26.21068	737	543 169	27.14774
638	407 044	25.25866	688	473 344	26.22975	738	544 644	27.16616
639	408 321	25.27845	689	474 721	26.24881	739	546 121	27.18455
640	409 600	25.29822	690	476 100	26.26785	740	547 600	27.20294
641	410 881	25.31798	691	477 481	26.28688	741	549 081	27.22132
642	412 164	25.33772	692	478 864	26.30589	742	550 564	27.23968
643	413 449	25.35744	693	480 249	26.32489	743	552 049	27.25803
644	414 736	25.37716	694	481 636	26.34388	744	553 536	27.27636
645	416 025	25.39685	695	483 025	26.36285	745	555 025	27.29469
646	417 319	25.41653	696	484 416	26.38181	746	556 516	27.31300
647	418 609	25.43619	697	485 809	26.40076	747	558 009	27.33130
648	419 904	25.45584	698	487 204	26.41969	748	559 504	27.34959
649	421 201	25.47510	699	488 601	26.43861	749	561 001	27.36786

TABLE VII. SQUARES AND SQUARE ROOTS (Continued)
1–1000

N	N^2	$\sqrt{N}$	N	N^2	$\sqrt{N}$	N	N^2	$\sqrt{N}$
750	562 500	27.38613	800	640 000	28.28427	850	722 500	29.15476
751	564 001	27.40438	801	641 601	28.30194	851	724 201	29.17190
752	565 504	27.42262	802	643 204	28.31960	852	725 904	29.18904
753	567 009	27.44085	803	644 809	28.33725	853	727 609	29.20616
754	568 516	27.45906	804	646 416	28.35489	854	729 316	29.22328
755	570 025	27.47726	805	648 025	28.37252	855	731 025	29.24038
756	571 536	27.49545	806	649 636	28.39014	856	732 736	29.25748
757	573 049	27.51363	807	651 249	28.40775	857	734 449	29.27456
758	574 564	27.53180	808	652 864	28.42534	858	736 164	29.29164
759	576 081	27.54995	809	654 481	28.44293	859	737 881	29.30870
760	577 600	27.56810	810	656 100	28.46050	860	739 600	29.32576
761	579 121	27.58623	811	657 721	28.47806	861	741 321	29.34280
762	580 644	27.60435	812	659 344	28.49561	862	743 044	29.35984
763	582 169	27.62245	813	660 969	28.51315	863	744 769	29.37686
764	583 696	27.64055	814	662 596	28.53069	864	746 496	29.39388
765	585 225	27.65863	815	664 225	28.54820	865	748 225	29.41088
766	586 756	27.67671	816	665 856	28.56571	866	749 956	29.42788
767	588 289	27.69476	817	667 489	28.58321	867	751 689	29.44486
768	589 824	27.71281	818	669 124	28.60070	868	753 424	29.46184
769	591 361	27.73085	819	670 761	28.61818	869	755 161	29.47881
770	592 900	27.74887	820	672 400	28.63564	870	756 900	29.49576
771	594 441	27.76689	821	674 041	28.65310	871	758 641	29.51271
772	595 984	27.78489	822	675 684	28.67054	872	760 384	29.52965
773	597 529	27.80288	823	677 329	28.68798	873	762 129	29.54657
774	599 076	27.82086	824	678 976	28.70540	874	763 876	29.56349
775	600 625	27.83882	825	680 625	28.72281	875	765 625	29.58040
776	602 176	27.85678	826	682 276	28.74022	876	767 376	29.59730
777	603 729	27.87472	827	683 929	28.75761	877	769 129	29.61419
778	605 284	27.89265	828	685 584	28.77499	878	770 884	29.63106
779	606 341	27.91057	829	687 241	28.79236	879	772 641	29.64793
780	608 400	27.92848	830	688 900	28.80972	880	774 400	29.66479
781	609 961	27.94638	831	690 561	28.82707	881	776 161	29.68164
782	611 524	27.96426	832	692 224	28.84441	882	777 924	29.69848
783	613 089	27.98214	833	693 889	28.86174	883	779 689	29.71532
784	614 656	28.00000	834	695 556	28.87906	884	781 456	29.73214
785	616 225	28.01785	835	697 225	28.89637	885	783 225	29.74895
786	617 796	28.03569	836	698 896	28.91366	886	784 996	29.76575
787	619 369	28.05352	837	700 569	28.93095	887	786 769	29.78255
788	620 944	28.07134	838	702 244	28.94823	888	788 544	29.79933
789	622 521	28.08914	839	703 921	28.96550	889	790 321	29.81610
790	624 100	28.10694	840	705 600	28.98275	890	792 100	29.83287
791	625 681	28.12472	841	707 281	29.00000	891	793 881	29.84962
792	627 264	28.14249	842	708 964	29.01724	892	795 664	29.86637
793	628 849	28.16026	843	710 649	29.03446	893	797 449	29.88341
794	630 436	28.17801	844	712 336	29.05168	894	799 236	29.89983
795	632 025	28.19574	845	714 025	29.06888	895	801 025	29.91655
796	633 616	28.21347	846	715 716	29.08608	896	802 816	29.93326
797	635 209	28.23119	847	717 409	29.10326	897	804 609	29.94996
798	636 804	28.24889	848	719 104	29.12044	898	806 404	29.96665
799	638 401	28.26659	849	720 801	29.13760	899	808 201	29.98333

TABLE VII. SQUARES AND SQUARE ROOTS (Continued)

1–1000

N	N^2	$\sqrt{N}$	N	N^2	$\sqrt{N}$
900	810 000	30.00000	950	902 500	30.82207
901	811 801	30.01666	951	904 401	30.83829
902	813 604	30.03331	952	906 304	30.85450
903	815 409	30.04996	953	908 209	30.87070
904	817 216	30.06659	954	910 116	30.88689
905	819 025	30.08322	955	912 025	30.90307
906	820 836	30.09983	956	913 936	30.91925
907	822 649	30.11644	957	915 849	30.93542
908	824 464	30.13304	958	917 764	30.95158
909	826 281	30.14963	959	919 681	30.96773
910	828 100	30.16621	960	921 600	30.98387
911	829 921	30.18278	961	923 521	31.00000
912	831 744	30.19934	962	925 444	31.01612
913	833 569	30.21589	963	927 369	31.03224
914	835 396	30.23243	964	929 296	31.04835
915	837 225	30.24897	965	931 225	31.06445
916	839 056	30.26549	966	933 156	31.08054
917	840 889	30.28201	967	935 089	31.09662
918	842 724	30.29851	968	937 024	31.11270
919	844 561	30.31501	969	938 961	31.12876
920	846 400	30.33150	970	940 900	31.14482
921	848 241	30.34798	971	942 841	31.16087
922	850 084	30.36445	972	944 784	31.17691
923	851 929	30.38092	973	946 729	31.19295
924	853 776	30.39737	974	948 676	31.20897
925	855 625	30.41381	975	950 625	31.22499
926	857 476	30.43025	976	952 576	31.24100
927	859 329	30.44667	977	954 529	31.25700
928	861 184	30.46309	978	956 484	31.27299
929	863 041	30.47950	979	958 441	31.28898
930	864 900	30.49590	980	960 400	31.30495
931	866 761	30.51229	981	962 361	31.32092
932	868 624	30.52868	982	964 324	31.33688
933	870 489	30.54505	983	966 289	31.35283
934	872 356	30.56141	984	968 256	31.36877
935	874 225	30.57777	985	970 225	31.38471
936	876 096	30.59412	986	972 196	31.40064
937	877 969	30.61046	987	974 169	31.41656
938	879 844	30.62679	988	976 144	31.43247
939	881 721	30.64311	989	978 121	31.44837
940	883 600	30.65942	990	980 100	31.46427
941	885 481	30.67572	991	982 081	31.48015
942	887 364	30.69202	992	984 064	31.49603
943	889 249	30.70831	993	986 049	31.51190
944	891 136	30.72458	994	988 036	31.52777
945	893 025	30.74085	995	990 025	31.54362
946	894 916	30.75711	996	992 016	31.55947
947	896 809	30.77337	997	994 009	31.57531
948	898 704	30.78961	998	996 004	31.59114
949	900 601	30.80584	999	998 001	31.60696
			1000	1 000 000	31.62278

selected answers to odd-numbered questions

CHAPTER 1

1. (a) 0, (c) 2,
   (e) -0.5, (g) -3
3. (a) $2x$, (c) $5x^4$,
   (e) $-4x^{-5}$, (g) $\frac{1}{2}x^{-1/2}$,
   (i) $\frac{1}{4}x^{-3/4}$, (k) $\dfrac{-2}{x^3}$,
   (m) nx^{n-1}, (o) 0,
   (q) $\frac{1}{2}x^{-1/2}$
5. (a) $6x + 2$, (c) $75x^2 + 32x + 6$,
   (e) $\frac{3}{4}x^{-3/4} + x^{-1/2}$,
   (h) $25x^4 + 24x^3 - 30x^2 + 32x + 25$
7. (a) $\dfrac{(x + 3) - (x - 2)}{(x + 3)^2}$, (c) $\dfrac{(x^2 + 2)(4x - 5) - (2x^2 - 5x)(2x)}{(x^2 + 2)^2}$,
   (e) $\dfrac{(x^4 + 3x^2)(-3x^{-4} - 2x^{-3}) - (x^{-3} + x^{-2})(4x^3 + 6x)}{(x^4 + 3x^2)^2}$
   (g) $\{(x^2 + 2x)[(x^2 + 4x)(3x^2 + 6x) + (2x + 4)(x^3 + 3x^2)] - (x^2 + 4x)(x^3 + 3x^2)(2x + 2)\}/(x^2 + 2x)^2$,
   (i) $\{(x + 3)(x^2 + 3x)[(x^2 + 5x)[2x(3x + 4) + 3(x^2)] + (3x + 4)(x^2)(2x + 5)] - [(x^2 + 5x)(3x + 4)(x^2)(x + 3)(2x + 3) + (x^2 + 3x)]\}/[(x + 3)(x^2 + 3x)]^2$
9. (a) 8, (c) 1,
   (e) $\frac{1}{8}$, (g) $\frac{3}{2}$,
   (i) 72
11. (a) $7 - x$, (c) $-5 + x$,
    (e) $10 - x$, (g) $1 - x$,
    (i) $-10 + \frac{7}{4}x$, (k) $\dfrac{5x}{4}$,
    (m) $\frac{3}{2} + \frac{1}{2}x$, (o) $10.75 + 0.75x$,
    (q) $\frac{93}{40} - \frac{3}{2}x$, (s) $69 + \frac{3}{4}x$,
    (u) $0.1 + 0.5x$, (w) $-19 + 5x$,
    (y) $-0.05 + 0.4x$

CHAPTER 2

1. (a) 3 minimum, (c) 3 minimum,
   (e) 3 minimum, 2 maximum, (g) -1 maximum, 1 minimum,
   (i) -1.37 maximum, 4.37 minimum, (k) $\frac{2}{3}$ minimum, -3 maximum

3.

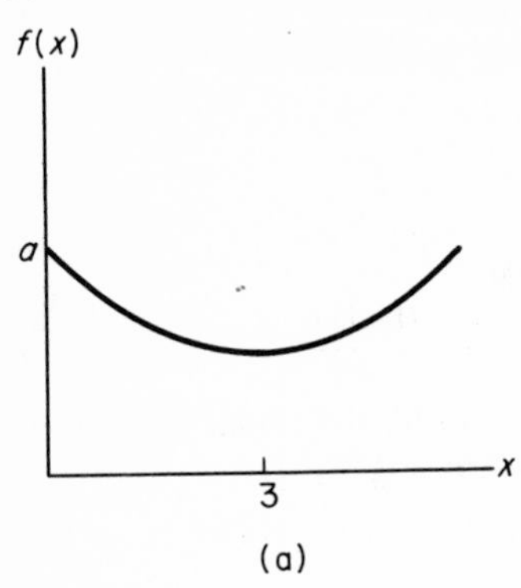

(a)

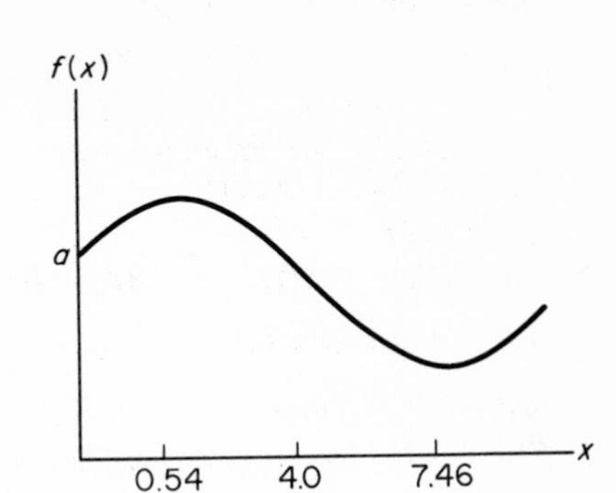

(c)

f(x)
a
x
−11.74 −6 −0.26

(e)

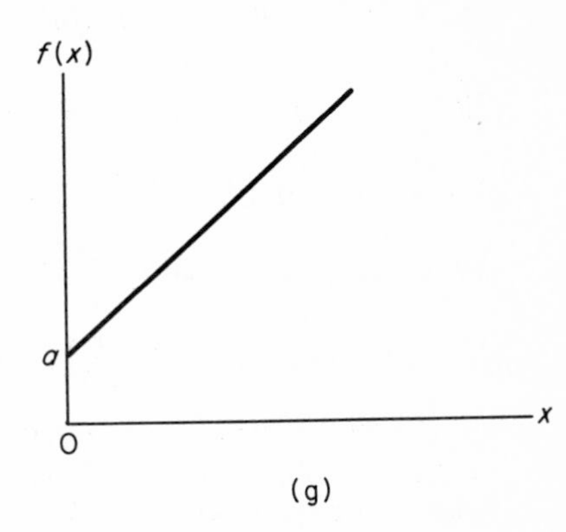

(g)

5. $c(x) = 5.9 - 0.0145x + 0.00001x^2$
   $c'(x) = -1.45 + 0.002x = 0$; $\therefore x = 725$ for minimum average cost.

7. $c(x) = 533 - 3.30x + 0.0067x^2$
   $c'(x) = -3.30 + 0.0134x = 0$; $\therefore x = 246$th unit for minimum average cost.

9. $p(x) = -3910 + 105.5x - 0.534x^2$
   $p'(x) = 105.5 - 1.068x = 0$; $\therefore x = 99$ units for maximum profits.

CHAPTER 3

1. $TR = 5.00q - 0.75q^2$
   $MR = 5.00 - 1.50q$
   $AR = 5.00 - 0.75q$

   $$E = \frac{1}{-0.75} \cdot \frac{5.00 - 0.75q}{q}$$

   Revenue is maximum when $q = 3.33$.

3. $TR = 2500q - 60q^2 - 0.5q^3$
   $MR = 2500 - 120q - 1.5q^2$

$AR = 2500 - 60q - 0.5q^2$

$$E = \frac{1}{-60 - q} \cdot \frac{(2500 - 60q - 0.5q^2)}{q}$$

$q = 17.2$ for maximum revenue

$TR(q = 17.2) = 20{,}303$

$AR = 1321, \; E = -1$

5. $p = 10 - 0.005q$ $\qquad AR = 10 - 0.005q$

$TR = 10q - 0.005q^2$ $\qquad E = \dfrac{1}{-0.005} \cdot \dfrac{(10 - 0.005q)}{q}$

$MR = 10 - 0.01q$ $\qquad q = 1000$ and $p = \$5$ for maximum revenue

7. $E = -1.20$; fare increase should not be made since $|E| > 1$.

9. $TC = 1000 + 20q - q^2 + 0.05q^3$ $\qquad MC = 20 - 2q + 0.15q^2$

$$AC = \frac{1000}{q} + 20 - q + 0.05q^2$$

$AVC = 20 - q + 0.05q^2$ $\qquad AFC = \dfrac{1000}{q}$

AC is a minimum when $\dfrac{d(AC)}{dq} = 0, q = 25.5.$

AVC is a minimum when $\dfrac{d(AVC)}{dq} MC = 0, q = 10.0.$

MC is a minimum when $\dfrac{d(MC)}{dq} = 0, q = 6.67.$

AFC is a minimum at the end point of the domain, $q = 30$.

11. $\pi = 600q - 0.5q^2 - 1500 - 150q + 4q^2 - 0.5q^3$

Profits are maximum for $q = 19.8$.

Revenue is maximum for $q = 600.0$.

AC is minimum for $q = 13$.

13. $TR = 200q - q^2$ $\quad TC = 5000 + 0.50q$

$\pi = 200q - q^2 - 5000 - 0.50q$

$q = 99.75$

15. $q = 1414$

17. $q_1 = 3830, \; q_2 = 8250$

CHAPTER 4

1. (a) $\dfrac{df}{dx} = 2 + 3y$ $\qquad \dfrac{df}{dy} = 3x + 3$ $\qquad \dfrac{d^2f}{dx^2} = 0$

$\dfrac{d^2f}{dy^2} = 0$ $\qquad \dfrac{d^2f}{dx\,dy} = 3$

(c) $\dfrac{df}{dx} = 6x(x^2 + y^2)^2$ $\qquad \dfrac{df}{dy} = 6y(x^2 + y^2)^2$

$$\frac{d^2f}{dx^2} = 6(x^2 + y^2)^2 + 24x^2(x^2 + y^2)$$

$$\frac{d^2f}{dy^2} = 6(x^2 + y^2)^2 + 24y^2(x^2 + y^2) \qquad \frac{d^2f}{dx\,dy} = 24xy(x^2 + y^2)$$

(e) $\dfrac{df}{dx} = 3(x + y)^2 \qquad \dfrac{df}{dy} = 3(x + y)^2 \qquad \dfrac{d^2f}{dx^2} = 6(x + y)$

$$\frac{d^2f}{dy^2} = 6(x + y) \qquad \frac{d^2f}{dx\,dy} = 6(x + y)$$

(g) $\dfrac{df}{dx} = \dfrac{1}{y} \qquad \dfrac{df}{dy} = \dfrac{x}{y^2} \qquad \dfrac{d^2f}{dx^2} = 0 \qquad \dfrac{d^2f}{dy^2} = \dfrac{2x}{y^3} \qquad \dfrac{d^2f}{dx\,dy} = -\dfrac{1}{y^2}$

(i) $\dfrac{df}{dx} = \dfrac{4}{2x + y} \qquad \dfrac{df}{dy} = \dfrac{2}{2x + y} \qquad \dfrac{d^2f}{dx^2} = \dfrac{-8}{(2x + y)^2}$

$$\frac{d^2f}{dy^2} = \frac{-2}{(2x + y)^2} \qquad \frac{d^2f}{dx\,dy} = \frac{-4}{(2x + y)^2}$$

(k) $\dfrac{df}{dx} = \dfrac{x + 2y}{x + 3y} + \ln(x + 3y) \qquad \dfrac{df}{dy} = \dfrac{3x + 6y}{x + 3y} + 2\ln(x + 3y)$

$$\frac{d^2f}{dx^2} = \frac{y}{(x + 3y)^2} + \frac{1}{(x + 3y)} \qquad \frac{d^2f}{dy^2} = \frac{-3x}{(x + 3y)^2} + \frac{1}{(x + 3y)}$$

$$\frac{d^2f}{dx\,dy} = \frac{-x}{(x + 3y)^2} + \frac{3}{x + 3y}$$

(m) $\dfrac{df}{dx} = 2xe^{x^2+y^2} \qquad \dfrac{df}{dy} = 2ye^{x^2+y^2} \qquad \dfrac{d^2f}{dx^2} = 2e^{x^2+y^2}(2x + 1)$

$$\frac{d^2f}{dy^2} = 2e^{x^2+y^2}(2y + 1) \qquad \frac{d^2f}{dx\,dy} = 4xye^{x^2+y^2}$$

(o) All derivatives are zero.

(q) $\dfrac{df}{dx} = 2x + y + yz \qquad \dfrac{df}{dy} = x + 2y + xz + z^2$

$$\frac{df}{dz} = xy + 2yz \qquad \frac{d^2f}{dx^2} = 3 \qquad \frac{d^2f}{dy^2} = 2 \qquad \frac{d^2f}{dz^2} = 2y$$

(s) $\dfrac{df}{dx} = \dfrac{1}{x + y + z} \qquad \dfrac{df}{dy} = \dfrac{1}{x + y + z} \qquad \dfrac{df}{dz} = \dfrac{1}{x + y + z}$

$$\frac{d^2f}{dx^2} = \frac{-1}{(x + y + z)^2} \qquad \frac{d^2f}{dy^2} = \frac{-1}{(x + y + z)^2} \qquad \frac{d^2f}{dz^2} = \frac{-1}{(x + y + z)^2}$$

(u) $\dfrac{df}{dx} = 2e^{(2x+y+z)} \qquad \dfrac{df}{dy} = e^{(2x+y+z)} \qquad \dfrac{df}{dz} = e^{(2x+y+z)}$

$$\frac{d^2f}{dx^2} = 4e^{(2x+y+z)} \qquad \frac{d^2f}{dy^2} = e^{(2x+y+z)} \qquad \frac{d^2f}{dz^2} = e^{(2x+y+z)}$$

3. (a) $x = -43.2,\ y = 33.6,\ \lambda = 230.4$
$f(-43.2, 33.6) = -2754.8,\ f(-42.2, 33.1) = -2753.5$
$f(-44.2, 34.1) = -2753.5$. Constrained maximum.

(c) $x = 5,\ y = 3,\ \lambda = -7$
$f(5, 3) = 28,\ f(4, 4) = 32,\ f(6, 2) = 32$. Constrained minimum.

(e) $x = 5,\ y = 4,\ \lambda = 2$
$f(5, 4) = -19,\ f(5\frac{1}{3}, 3) = -25\frac{1}{3},\ f(4\frac{2}{3}, 5) = -25\frac{1}{3}$
Constrained maximum.

(g) $x = 7.45,\ y = 4.55,\ \lambda = -10.8$
$f(7.45, 4.55) = 98,\ f(7, 5) = 60,\ f(8, 4) = 48$
Constrained maximum.

5. $x = 10.68,\ y = 42.72,\ c(10.68, 42.72) = 34{,}430$
7. An increase in price of \$0.01 results in a decrease in quantity of 0.3015 unit. An increase in advertising from \$10 to \$11 results in an increase in quantity of 0.4067 unit.
9. $x = 17.73,\ y = 78.26,\ \lambda = -0.057$
$p(17.73, 78.26) = 38{,}057$

CHAPTER 5

1. $q = 121,\ s = 66,\ t = 0.242$
3. $q = 667,\ t_1 = 0.0333,\ t_2 = 0.0500,$
$t_3 = 0.1500,\ t_4 = 0.1000,\ t = 0.3333$
Peak inventory = 300
5. $q = 20.25 \simeq 20,\ s = 14,$
$t = \frac{2}{3},\ t_2 = \frac{1}{5}$ month
9. $i = 1.12\%$
11. $i = 4.96 + 4.70 = 9.66\%$
13. $i_{\text{daily}} = 5.13\%,\ i_{\text{qtrly.}} = 5.09\%$
15. 8% for alternative 1
8.10% for alternative 2
17. $p = \$2665$

CHAPTER 6

1. (a) $6x + c$ (c) $\dfrac{x}{3} + c$ (e) $ax + bx + c$

(g) $x + c$ (i) $\dfrac{x^2}{2} + c$ (k) $-x^{-1} + c$

(m) $-\frac{1}{2}x^{-2} + c$ (o) $\dfrac{2x^{3/2}}{3} + c$ (q) $\frac{8}{3}$

(s) 5 (u) $7499\frac{1}{4}$

2. (a) $\dfrac{3x^2}{2} + 4x + c$ (c) $\dfrac{x^4}{2} + \dfrac{x^3}{3} + c$

(e) $\dfrac{x^3}{3} + x^2 + x + c$ (g) 15

(i) $e^{3x} + c$ (k) $\dfrac{e^{2x}}{4} + c$

(m) 1 (o) $x \ln(x) - x + c$

(q) $x \log_a(3x) - x \log_a e + c$ (s) 14.0260

3. (a) $\frac{2}{7}(x-1)^{7/2}+c$ (c) $\dfrac{(4x^3-2)^{10}}{120}+c$

(e) $\dfrac{e^{x^3}}{3}+c$ (g) $\frac{1}{5}\ln(5x-6)+c$

(i) 2.4423, (k) $\frac{1}{2}(x^2-6)\ln(x^2-6)-\frac{1}{2}(x^2-6)+c$

(m) 0.6932

CHAPTER 7

1. (a) 36, (c) 132, (e) 128, (g) 21.25,
   (i) -7.5, (k) 30, (m) $-41\frac{2}{3}$, (o) 54,
   (q) 2893, (s) $-87\frac{1}{12}$, (u) $\frac{2}{3}(e^{60}-1)$, (w) 0.393,
   (y) 0.234
3. (a) $k=\frac{1}{60}$, (c) $k=\frac{2}{9}$, (e) $k=\frac{2}{49}$, (g) $k=\frac{2}{27}$,
   (i) $k=\frac{1}{126}$, (k) $k=3$, (m) $k=\frac{3}{190}$
4. (a) 5, (c) 1, (e) 7.3, (g) 3.17,
   (i) 4.28, (k) $\frac{1}{3}$, (m) 2.26
5. (a) 8.3, (c) 0.5, (e) 8.21, (g) 0.38,
   (i) 1.75, (k) $\frac{1}{9}$, (m) 0.30
6. (a) 5, (c) 0.88, (e) 7.95, (g) 3.3,
   (i) 4.57, (k) 0.23, (m) 2.34
7. (a) 0.25, (c) 0.607, (e) $\frac{3}{10}$

index